TREASURES TO BE FOUND IN ALL FIFTY STATES!

Maine: Captain Kidd buried treasure all over the state—so did many others!

Massachusetts: The Brink's Robbery loot has never been found, and neither have dozens of pirate riches.

North Carolina: This state, as with the other southern states, is loaded with treasures from pirate loot to Civil War gold, and more!

Texas: The Lone Star State, with its colorful history of bandits, barons, and buccaneers, has more hidden treasure than any other state.

Montana: Before Montana re-nicknamed itself the Big Sky Country, it was known as the Treasure State.

Colorado: The Rocky Mountains held a treasure trove of gold mines, and now has more ''lost'' mines than any other state except California.

Hawaii: With a royal history comes the intrigue of legendary caches of gold hidden all over the island.

AND MORE . . . MUCH MORE!

HIDDEN TREASURE

WHERE TO FIND IT, HOW TO GET IT

BILL YENNE

AVON BOOKS  NEW YORK

AVON BOOKS
A division of
The Hearst Corporation
1350 Avenue of the Americas
New York, New York 10019

First Avon Books Printing: May 1997

AVON TRADEMARK REG. U.S. PAT. OFF. AND IN OTHER COUNTRIES, MARCA REGISTRADA, HECHO EN U.S.A.

Printed in the U.S.A.

RAI   10 9 8 7 6 5 4 3 2 1

CONTENTS

6. THE TREASURES OF THE NORTHEAST   48

7. THE TREASURES OF THE SOUTH   69

8. THE TREASURES OF THE GREAT LAKES STATES   124

INTRODUCTION

Tales of buried treasure and lost mines are interwoven throughout the fabric of history, and treasure seekers themselves been responsible for some of history's most important milestones. Indeed, the original European obsession with America itself began not so much with a longing for adventure or a taste for spices, but with a lust for gold. The hundreds of thousands of people who abandoned jobs, lifestyles, and families to scramble to California in 1849 were not coming for sun, sand, and surf.

Meanwhile, treasure seekers of another sort—including historians and archaeologists—have been responsible for the discovery and excavation of the sites and cultural treasures that help us know and understand our history.

This book specifically tells of locations where one *might* find treasures of monetary value, but in searching for nearly all of these sites one *will* find stories of the people who helped to form our history, whether they are legendary or notorious household names, or simply long-forgotten characters whose dreams of riches led them to a place in a dusty historical footnote.

The story of the human obsession with treasure, especially with gold, dates to the deepest prehistoric dawn of human civilization and continues to this day. It is a cold, straw-colored metal that ignites a fire in the soul.

Christopher Columbus himself wrote, ''Gold is the most excellent thing. Of gold, treasure is made, and he who possesses gold does what he wishes in the world.''

The value of gold is in its rarity. It's axiomatic that anything worth having is not easy to get. Nearly five centuries after the lucky Chris stood on the sands of San Salvador, Berwick ''B. Traven'' Torsvan wrote in *The Treasure of the Sierra Madre* (which John Huston turned into the legendary 1948 film of the same name with Humphrey Bogart), ''You thought the gold would be lying around like pebblestones, and nothing to do but bend down and pick it up and go off with it by the sackful. But if it was as simple as all that, gold wouldn't be worth more than pebblestones.''

Columbus and Traven present two tangential views of treasure, that is that it is a wondrous thing to have, and that it would not be worth having if it was easy to have. Traven also implies that the search for the treasure is as much a part of the experience as having it. Neither Columbus nor Bogart's character in *The Treasure of the Sierra Madre* died rich or happy.

For Columbus and his Spanish employers, gold was easy to find at first. They took it the old-fashioned way—they stole it. The easy gold was found and carted off to Europe or buried again to be searched for by other treasure seekers many centuries later. Since the Spanish era, numerous people have searched the Western Hemisphere for gold. Many have died trying, but some have become fabulously wealthy. In Traven's book, the gold came easy, but the psychological baggage that came with it proved to be an impossible burden.

In 1848, James Marshall found a gold nugget in California's Sierra Nevada and set off a spark of excitement that created the biggest voluntary mass migration in modern history and one that literally created the state of California.

Billions of dollars were made in the gold fields of California, Arizona, and Colorado as well as in the silver mines of Nevada's Comstock Lode, but there are billions left undiscovered. There are lost mines and buried stashes of old Spanish gold throughout the American West.

As Traven clearly conveyed in the *Treasure of the Sierra Madre*, and as countless others have conveyed in other tales of treasure, the search inflames the mind and fires the spirit, but it is a wildfire that can flame beyond control. A word or two whispered about a lost mine or a buried strong box will turn the most mild-mannered, most settled person into a Walter Mitty or a Jesse James.

The treasure lore of the United States is an essential element of our history, and one of the most adventurous aspects of our history. It was treasure that inspired people to explore, to take chances, and to build the infrastructure that made the lifestyles of both Mitty and James possible.

The treasure lore of the United States is a cornucopia of lost Spanish mines, of gold and silver stolen from Mexican pack trains, hidden by Native Americans, stolen from churches, concealed by priests, and whispered about by paranoids with a past. It includes mines lost and found by Frenchmen, Dutchmen who were really Germans (''Deutsche'' men), Dutchmen who were really Dutchmen, one-eyed strangers, English pirates, and Spanish conquistadors. The stories tell of silver, gold, precious stones, and confederate coins stolen from army convoys, cruel slavers, innocent travellers, guilty travellers, and people who disappeared without a trace.

A generation or two ago, stories of lost Indian or lost Spanish mines were still fresh in the minds of the loners and packers that lived in the hills of the West, or hard-to-reach villages on the Outer Banks, and these legends became a staple of pulp journalism. Tales were often related around flickering camp fires to the eager ears of greenhorns by old Mexican men who were often part Indian or who had been raised by the Indians. Almost invariably, these old men were sure that for the right inducement, they could find that mine again. Occasionally— not often, but often enough—they actually did.

Today, the old-timers who were adults in the early twentieth century are gone, taking their tales with them. Many of the more enduring legends were recorded for posterity, but a rich literature has grown obscure, more the stuff of dusty shelves of Tucson, Idaho Falls, or Helena used bookstores than of current tabloid journalism.

Many of America's treasures are associated with the names of history's notorious rascals. On the eastern seaboard and the Gulf Coast, you can practically tell where you are by the name of the buccaneer that falls from the lips of anyone describing pirate treasure. If they whisper ''Captain Kidd,'' you're probably north of the mouth of the Delaware River, and if the name is "Blackbeard," then you are farther south. On the Gulf Coast, if someone says "Billy Bowlegs," you may well be in the Florida panhandle, but if it's "Jean Lafitte," you could be anywhere from Mobile to South Padre Island. While many of the treasures identified with these gentlemen may actually be tied (for local publicity purposes, of course) to loot stashed by pirates of lesser notoriety, each of these villains was extremely active in their respective areas. In the case of Jean Lafitte, some people say he never left, that his ghost still wanders through the bayou country west of the Mississippi Delta.

Out west, there are stories told from Texas to the Dakotas of blood-drenched loot hidden by Jesse and Frank James and their gang. In the Rockies, tales turn to Butch Cassidy and his Wild Bunch, and in Montana, they speak of Henry Plummer, who conducted his reign of terror from behind a badge and died with his boots on without divulging the whereabouts of a dozen supposed caches.

In California, where the '49ers grossed a billion dollars a year for the better part of a decade and turned up one single "nugget" weighing 195 pounds, people still earn a good living panning for gold, and the hidden treasure stories—if you believe them all—contain enough value to pay off several national debts.

The Southwest, with its wind-swept mesas and labyrinthine canyons, has beckoned to treasure-seekers for centuries. Ever since 1536, when Cabeza de Vaca stumbled, half dead, into Mexico City with tales of "seven golden cities," people have come into the Southwest to pursue treasures of mythical proportions. It is a place of real and well-documented gold and silver mines, as well as tall tales of stolen coins and fields of gold nuggets "the size of quail's eggs" bragged about by men in clapboard saloons, men who had not tasted whisky for eighteen months, and by men who would never again see the sun rise.

Treasure is elusive in a metaphysical as well as a physical sense. There is a story that they used to tell in the mountains of Grant County, New Mexico that is a good allegory for this. It is the tale of "Los Perros de la Niebla"—the Dogs of the Mist. It was nearing winter, the air was cold, the canyons were filled with mist, and you could smell the snow coming. A man and his dogs were deep in the mountains, hunting for a bear, when he stumbled across a stream bed filled with large gold nuggets. Somehow, in his excitement, his gun apparently went off, mortally wounding him. A day or so later, the dogs were seen in town and people wondered what had happened to their master. Around the neck of one dog was a torn piece of the man's shirt. On it, written in blood, were the words "Follow Dogs. I am Dying." Tied up in the cloth was a huge nugget of almost pure gold.

Unlike the happy ending you'd expect if the story originated in Hollywood, searchers could not get the dogs to lead them back to the man. They searched the misty mountains—probably until the snows finally came—but they never found the man, and they never found the gold. However, as the story goes, on misty days in those mountains, you can still hear the distant barking of "Los Perros de la Niebla."

Treasure hunting is indeed like following dogs in the mist. There is a tangible story and just enough of a hint of location to keep us going. There is always something intangible that drives the treasure hunter, something more than the thirst for money. The odds are probably better with lottery tickets. Lottery tickets are certainly less work.

Treasure is magic, a magic that transcends the monetary value of the gold or silver itself, a magic that comes from the legends that swirl about the treasure, and a magic that is part and parcel of the search itself, which often is transformed into a search of self-discovery, and the search itself becomes the ultimate treasure.

The ethos and mythos of the treasure hunter is probably best summarized by Frank Dobie, a man who wrote a lot about the treasures of Mexico and the Southwest in the 1920s. Wrote Dobie, "What men believe or fancy to be true, what they have faith in, whether phantom or fact, propels their actions. The hunter of precious metals is always a fatalist, no matter how civilized above superstition he may be. Deep in his heart he believes that somewhere out in the sierras, the magic scales are awaiting him."

We wish you good luck, but more than that, we wish you a wonderful adventure on the treasure trails of a fascinating land.

1

USING THIS BOOK
TO SEARCH FOR TREASURE

After everything else is said and done, this book is about treasure and how to find it. As with any endeavor, there are no guarantees of ultimate success, only guarantees of failure without proper planning. You should begin with the expectation that your search will be as much a search for adventure as for treasure.

Legends and tales of treasure are just that. If the details about these treasure sites were specific, and if the gold was, as Traven wrote in *The Treasure of the Sierra Madre*, "lying around like pebblestones, and nothing to do but bend down and pick it up and go off with it by the sackful," someone already would have been off with it. It's not that simple. People have been searching for and dreaming of some of these treasures for years. It will take time and patience, and it will take care and research, but in that time and research you will find that the true treasure may very well be in the experience of the quest.

Before you do anything else, read through each of these steps and be sure that you are prepared to properly plan your treasure hunt.

To facilitate your search and make it as expeditious as possible, potential treasure sites are grouped and discussed by county. For users who are searching for treasure, there is an economy of scale in looking for several related sites in the same area. In many cases, sites within a county or adjacent counties may be related to one another by way of a common mining area, the activity of a particular pirate or outlaw gang or some geographical feature that made a place in the county a particularly good place to hide treasure. In some cases, two or more may be variations that evolved from a single story, and in other cases stories may have evolved from deliberate deceptions that were concocted a century ago by a paranoid prospector who wanted to divert greedy rivals from his claim.

Legends of treasure are typically unclear or ambiguous about specific locations. It is understandable that the more specific the directions to the site, the more likely that the treasure (if it ever really existed) has already been found. When we use words such as "in," we mean to imply very close proximity. When we say that a site is "near" a named town or geographic feature, it means that it is closer to that town or place than any other of similar size or importance and probably within a five-mile radius in densely populated states and a ten-mile, twenty-mile, or greater radius in more sparsely populated areas where towns are farther apart.

This book is only a point of departure. Always make discreet inquiries locally.

The dollar figures given for the value of particular caches have not been adjusted to current dollar values, because most of these figures are part of the original legend and are obviously subject to exaggeration in the telling or retelling. They might also be based on rough estimates made hurriedly on the spot.

Estimates associated with chests or bags of stolen loot will have some basis in reality, but rough guesses at the potential future value of mines or placers are almost invariably exaggerations for effect, or underestimates for the purpose of security. For this reason, they are universally unreliable and misleading, so we have omitted them. For example, there are several treasures that are valued at "$10 million," which may actually be worth $100 million or $10,000. If one believes the nineteenth-century valuation estimate of Arizona's legendary Lost Dutchman Mine, it would now be worth in excess of $7 billion.

A dollar value associated with a legend may have originated

in the 1870s and have been adjusted for "current" value in the 1940s, or not adjusted at all. To adjust the figure today without knowing whether it had been previously adjusted would be very misleading. For example, the value of the metal in a twenty-dollar gold piece is now worth over four hundred dollars, so an estimate based on the face value of such coins must be adjusted upward by a factor of five. We have included such figures wherever they are part of a given legend, and not as a guarantee of value.

Locations too, may be variable. In some cases, legends are based on place names that may have been in use in the eighteenth or nineteenth century, but which are no longer current. If a specific name (or a similar name) cannot be found on current maps, it is good to inquire locally, asking whether anyone remembers a place that *used* to be known by such a name, or if there is a similar name in an adjacent area.

FOUR STEPS TO TAKE AT LEAST SIX TO EIGHT WEEKS BEFORE LEAVING HOME

Step One

You should start by identifying your area of search. You could start near where you live, or near where you have already planned to go on vacation. You may also read the listings and pick a place because the stories of the treasure in that area fire your imagination. In order to make the best use of your time, it's good to focus on a specific county or a cluster of adjacent counties, and to pick several possible treasure sites in that area. The book is organized by county in order to facilitate your search.

Step Two

When you have picked a general area, obtain a simple gas station or auto club road map of the state or states where the area is located. Nearly every town, road, mountain range and river mentioned in this book as a landmark will be on a standard road map or highway atlas. Using this book and your road map, you should be able to mark the general location of the sites within your specific county or cluster of adjacent counties.

Step Three

Having narrowed your geographic scope, it's time to get your hands on the most detailed map possible. The United States Geological Survey (USGS) publishes highly detailed 1: 24,000 scale topographic maps and more general 1:100,000 scale maps of the entire United States.

To give you an idea of how detailed these maps are, the state of Arizona takes up a page and a half in a typical highway atlas. This map shows all paved and many unpaved roads in perfect clarity, and it shows towns from Phoenix down to those with populations of less than fifty. In this map, the state is about fourteen inches wide. If one used a set of USGS 1: 24,000 maps to form a map of Arizona, the state would be more than eighty *feet* wide!

USGS topographic, or "topo" maps, are incredibly detailed and are among the best maps published anywhere in the world. Your road map may show towns with populations of less than fifty, but a topo map will show individual buildings within the town, windmills on the edge of town and the contours of small hills and gullies, as well as dry stream beds and ponds the size of a large living room and dirt roads that you might miss driving past.

You should not plan or undertake any actual field work without a topo map, and you should have your topo map in your possession at all times while you are in the field.

To get the correct topographic map or maps, you should request the *Index to Topographic and Other Map Coverage* for the state or states that you desire. These publications are available for free from the USGS and are a guide to all the topo maps available for the state. Using this guide, you may purchase the map or maps that you wish for a nominal fee.

They are inexpensive, yet they are the single most important tool that you can have. Spend the extra money and buy maps of adjacent areas if there is any possible chance that you will be working in that area, or that you will have to drive or hike through an adjacent area to get from a main road to your goal.

Obtaining the index and purchasing the maps may be done in person at the regional USGS center in your area (consult your white pages under the US Government section), or by contacting

The US Geological Survey
Map Distribution
USGS Map Sales, Building 810
Denver, Colorado 80225
(303) 235-5829

Step Four

At this point, it is usually a good idea to go to your library and/or bookstore and read up on the history of the region, and if possible, the historic period when the treasure was hidden, or when the mines were in operation. For example, if it was hidden during the Civil War, you should read about Civil War battles and campaigns that took place in the area. If the treasure was hidden by a specific famous pirate or outlaw, such as Captain Kidd or Jesse James, reading a biography of that person will help give you some background information that may be vital later in your search.

Gold mining activity in a specific area typically was confined to a short time span, and histories of many of these "booms" have been written and are available in bookstores or libraries. Back issues of regional newspapers are also valuable, and can be used to clarify specific minutia. In some cases, entire books have been written about specific treasures.

In short, you should know as much as possible about the circumstances behind the treasures that you're seeking.

Two Steps to Take When You've Arrived in the Target County

Step One

Having arrived on site with your topo maps and your overview of the history of the area and the treasure, you should continue what you started in Step Four above. No matter how good your library was at home, local libraries are always the best place to flesh out your basic knowledge of an area and its history. Most libraries have a local history section, and many libraries and county historical museums are staffed by knowledgeable people who love to answer questions about local history and local lore. They may know—or may be able to refer you to people who know—more about a specific event or treasure than has ever been written in a book.

Your research may take you to several libraries and/or museums. Often the most comprehensive information will be in the county seat, or in the largest town in the area, which may be across the county, or even the state, line.

If a treasure is associated with a specific event, a look in back issues of the local or county newspaper should reveal helpful details. Information for narrowing down or pinpointing the date of an event may be found in one of your books, or in the memory of someone at the library.

If a treasure is associated with land that was once owned by a specific individual, a trip to the county land office may be in order. Land offices are also a wealth of information and referrals about mining claims in the area, and may direct you to treasure sites that are not included in this book.

If the treasure is located in or near a mark or other designated site, the visitor's center or information office may have helpful information.

Step Two

Using the information that you have obtained by now, you should be able to pinpoint an area of search on your topo map. You may even know exactly where you want to go.

FOUR STEPS TO CONSIDER BEFORE GOING TO THE SPECIFIC SITE

Step One

Don't trespass. Carefully reread Chapter 3. Private property is private. Public land is available for public use, but there are important rules and restrictions.

Step Two

Be careful. Carefully reread Chapter 2.

Step Three

Fill your gas tank and your canteen. Make sure that you have your topo map, and any notes you've made, in your possession.

Step Four

Be prepared. The old Boy Scout motto applies. Don't proceed unless you are sure that you are acting legally and that you are not endangering your health or safety. Having done this, you should be sure that you have the necessary provisions and the necessary tools.

You may be able to drive to the site in an ordinary car, or you may need an off-road vehicle. You may have to hike for an hour or several days. In each case, you should be absolutely sure that you are capable of making the trip and that you have all the food, water, and gas that you need for a *round trip* and for the length of time you plan to spend at the site.

Before you leave, you should create a realistic checklist of things that you will need. Essentials for any search, even if most of the trip to the site is in a vehicle, are

- your topo map
- water bottle or canteen
- a reliable compass
- a reliable watch
- flashlight (never a priority at 10 am, but often a vital necessity eight hours later)
- an extra flashlight and *two* sets of extra sets of batteries (if you will be entering a cave)
- first-aid kit
- matches or lighter
- proper footwear

- proper hat and coat, considering temperature and possibility of rain or snow
- sunscreen or sun block
- gloves
- day pack or larger pack to carry the other items, unless you'll be within five to ten minutes of your vehicle
- shovel, pick, gold pan, and/or metal detector
- rope (usually not necessary, but often useful)
- hacksaw blade (for removing the padlock from the strong box).

If you are going to be spending the night outdoors, you should be prepared by knowing how to camp. Often, some advice from people at a camping supply store and one or two nights at a drive-in campground are all that it takes, but information and experience are essential. In addition to the items listed above, you should take

- a sleeping bag and (if desired) a pillow
- a lightweight tent that you can carry easily
- insect repellent (even if invisible in the daytime, they come out of nowhere at sundown)
- food that can be prepared at the campsite
- a camp stove and food-serving utensils
- extra plastic bags
- water purification tablets
- toiletries (including soap and toothpaste) and toilet paper.

USING A METAL DETECTOR

Metal detectors have revolutionized the search for lost and buried treasure. They originated a half century ago as a military tool for use in finding buried land mines, and were eventually adapted for civilian use in searching for buried metallic objects such as coins and artifacts.

Modern metal detectors are quite rugged and sophisticated instruments. They weigh between two and five pounds and can be operated easily with one hand. Some can be adapted for mounting to your belt. Depending on type, metal detectors can be used to find metallic objects buried as deep as twelve inches or more. They can also be adjusted to discriminate between objects. For example, many modern detectors can be programmed to reject aluminum can pull-tabs and bottle caps, while reporting coins. Many metal detectors are waterproof, and some are designed for use under water.

The average price for a good metal detector is between $400 and $600, but there are both cheaper and more advanced models. As with purchasing any other tool or piece of equipment, you should investigate the detectors with the features that you need that are available in your price range, and do some comparison shopping.

After you have compared and selected a metal detector—and before going into the field—you should practice with it in a controlled situation. Start with something as simple as having a friend bury some pocket change in your backyard. The next step would be to bury a variety of objects so that you could get the feel for how your detector reacts to bottle caps, nails, and larger metal objects.

A good publication to read before selecting a metal detector is *Metal Detector Information*, which is published by Tesoro Electronics, a leading maker of metal detectors. It contains information about Tesoro metal detectors, of course, but also a good deal of useful information about the selection and use of metal detectors in general, and a list of dealers throughout the United States who sell metal detectors and related equipment. For more information, contact

Tesoro Electronics Inc.
715 White Spar Road
Prescott, Arizona 86303
(520) 771-2646

DIGGING FOR THE TREASURE

How you "dig" for the treasure depends entirely on the nature of the treasure. Some treasures are actually buried.

Since many have been buried for over a century, clues have probably shifted or are gone. Floods and landslides may have buried the treasure deeper or even pushed it to a different place.

Placer gold can be found the old fashioned way—with a gold pan. You take a pan of gravel from a shallow place in the stream, swirl the water in it, allowing the gravel to gradually wash out. The heavier gold will work its way to the bottom and will be clearly visible when most of the gravel is gone.

Take care when digging, especially when digging or working on hillsides, with loose rock, or in caves and crevices. Do not enter abandoned mine shafts. There are few places that are more dangerous. There is no treasure worth taking a chance on getting caught in a mine shaft cave-in.

Treasure may not be buried. It may be out in the open in a place that is so remote that no one has been there in a hundred years. It may be disguised as something else.

Finding the treasure once you've reached what you think is the site takes a little intuition and a lot of patience and luck. Here, you're on your own. You have to put yourself in the place of the person who buried it, asking, "Where would *I* have put it?"

2

NOTES OF CAUTION

This book is intended as a guide for those wishing to visit and experience the sites of interesting events of American history. If you leave a designated trail or disturb the environment in any way you must be prepared to deal with and accept the consequences.

Thoroughly read Chapter 3 before you start. You may travel freely in most federal and state public lands. You may follow maps and look for and at anything you wish. However, in many cases, you may not dig for or remove natural features or artifacts without permission in the form of a permit.

The reader must bear the full responsibility for determining the ownership of the land, of understanding whatever restrictions there may be on various activities undertaken on that land, and for making necessary arrangements with the owner for access to nonpublic land. Do not cross fences without permission. Trespassing is a crime not taken lightly in most rural areas.

Many of the sites are on public land to which access is not restricted, but some are on private or Native American land which may not be posted. Some are on military bombing ranges where access is strictly forbidden and extremely dangerous.

We do not intend to recommend or advise that anyone do anything reckless or dangerous, or infringe on the property rights of others. Look, but don't touch without permission.

Many of the sites listed in this book are in wilderness areas where there are serious environmental dangers. Use utmost caution when travelling in the wilderness. When out of the sight of paved roads and familiar landmarks (on foot or in a vehicle), always carry a detailed topographic map and compass as well as adequate food and water. In areas where it is specifically required, stay on marked trails and/or roads.

Be wary of hazards from poisonous snakes, gila monsters, scorpions, tarantulas, mountain lions, and other predators. Even insects, especially ticks carrying Lyme disease or Rocky Mountain spotted fever, can be dangerous. Always carry insect repellent. In the northern Rockies of Idaho, Montana, and Wyoming, you may encounter grizzly bears, which are especially aggressive and dangerous. Always inquire locally regarding potential predators. Do not deliberately put yourself in danger.

When hiking in the wilderness, or when driving on unpaved roads or off-road areas, always carry a watch and allow plenty of time for your trip. Consider distance, elevation, weight being carried, physical condition (of yourself, your travelling companions, and your vehicle), weather, and hours of daylight.

Check the latest weather forecast, and be prepared for sudden changes in weather. Even in summer, cool, wet conditions, if accompanied by winds, can cause hypothermia. On warm, sunny days, even if it is not particularly "hot," people can be subject to heat exhaustion.

There are certain times of year that expeditions into wilderness areas simply should not be attempted. Hiking in snow-covered mountains, especially the Cascades, Rockies, and Sierra Nevada, should not be attempted in winter. Desert excursions in the summer, such as into California's Death Valley and the deserts of Arizona or New Mexico, where temperatures can stay well above 100 degrees from dawn to dusk, should not be attempted.

If you are going to be driving on an unpaved road or on an especially remote paved road, always start with a *full tank of gas* and carry both a spare tire and a flashlight. In winter, always carry chains if there is even the remotest possibility of snow or ice.

When exploring coastal areas, always be aware of waves, undertows, and tides. Each year, dozens of people are stranded by incoming tides when taking dangerous chances or not paying attention.

Also take care when digging or working with loose rock, or in caves and crevices. We strenuously advise against entering abandoned mine shafts!

Just don't do it.

Years of inattention, water seepage, the effects of freezing and thawing, and possible seismic activity have rendered virtually every abandoned mine in the country prone to imminent collapse, cave-in, or flooding. It is best to observe these from the entrance and speculate about caches left outside. They were usually abandoned when they played out, and there is almost certainly nothing of value inside them.

If you are going to be entering a naturally occurring cave or cavern, beware of loose, fallen, or falling rock. Do not enter such a cave if there is potential danger from such rock, or if you have claustrophobia or difficulty with directions and navigation. In caves and caverns, *always* carry an *extra* flashlight and *two* sets of extra sets of batteries. Also beware of predators, which range from poisonous insects and reptiles to bears.

Bat caves, which are common in Texas and the Southwest, are a hazard not for the bats themselves, which are almost certainly harmless, but for deadly disease germs often present in bat guano, which litters the floor of bat caves.

If there is any possibility of danger, stay out of the cave.

Remember, you must bear the full responsibility for determining the ownership of the land, of understanding whatever restrictions there may be on various activities undertaken on that land, and for making necessary arrangements with the owner for access to nonpublic land.

You also must bear the full responsibility for your own physical safety and that of your travelling companions. The odds of incurring physical injury in potentially dangerous situations are always much greater than the odds of finding buried treasure anywhere.

Leave nothing to chance. Always inquire locally about specific hazards and road conditions. Always pay attention to changes in tides and weather. Do nothing reckless or dangerous, or infringe on the property rights of others.

3

WHOSE TREASURE IS IT?

One of the most important questions to consider in the search for buried or hidden treasure is your legal right to take possession of it and keep it as your own property. While laws concerning the recovery of sunken treasure found beneath international waters in the world's oceans are quite liberal, laws governing treasure troves on land or in coastal waterways can be quite restrictive.

Broadly speaking, there are three types of land ownership in the United States: government-owned, privately owned, and foreign government–owned. The latter includes embassies and consulates of foreign countries which are legally considered to be the "soil" of the country in question. These are mostly very small. They include single buildings or small compounds, such as the embassies in Washington DC, or office buildings or even office suites, such as consulates in cities like New York and San Francisco, or the United Nations missions in New York City. Because of their small size, the unlikeliness of treasure on their premises, and the fact that they are not generally open to the public, we have chosen not to cover them.

Private land includes property owned by individuals (such as houses and condominiums), owned by families (such as farms or vacation homes), owned by companies (such as factories or farms), or owned by organizations (such as scout camps, churches, private schools).

Meanwhile, cities, counties, states, and the federal government all own, manage, and/or administer land. Some of this government-owned land is used for government offices, museums, and maintenance facilities. Much government land is set aside as parks for the recreational use of everyone. Some government areas, such as in the National Forest System or in the case of Bureau of Land Management grazing lands, are leased for commercial use. Most government land is theoretically "public," and as such has generally open access (sometimes through a nominal fee), but some government land, like military bases, is restricted and no public access is permitted.

People entering public land become subject to state or federal law as they cross into public land jurisdictions, whether or not the boundaries are fenced or marked. To avoid misunderstandings, we recommend contact with the local office having jurisdiction over the public lands prior to any searching or collecting.

It is important to note that entering upon or disturbing any publicly owned land will almost certainly be subject to rules and regulations of the agency that controls that property. When seeking permission from an agency to enter a specific area, consent given to enter will almost certainly contain limitations with respect to any property found on the land.

There will also be tax implications regarding the discovery of treasure, if it is determined that the finder owns or has any ownership interest in that property. You should contact the state department of taxation or your tax advisor in that regard.

People entering private land become subject to laws governing private property rights, whether or not the boundaries are fenced or marked. To avoid misunderstandings, we recommend contact with the local land office having jurisdiction in the area prior to any searching or collecting.

Discovery of human remains of any kind, whether or not they appear to be prehistoric, should be reported immediately to the local police or sheriff's office.

Since the passage of the Antiquities Act of 1906, it has been illegal to excavate or appropriate artifacts on public lands without permission. This would include hunting for treasures, artifacts, and gold in any form other than as found in its natural mineral condition. The reason is to preserve, enable study of, and understand our nation's heritage. Collecting artifacts from the surface or unsupervised digging is not a constructive way to participate. Unauthorized activities destroy potentially important archeological information about the context in which artifacts are found. Federal agencies are charged with managing historic and archeological resources for the benefit of the American public. The theory is that if these disappear, artifact by artifact, soon so many chapters to the story are missing that the story cannot be told.

FEDERALLY OWNED AND/OR ADMINISTERED LAND

In the eastern part of the United States, most land is in private hands, while in the West, specifically the Rocky Mountain states, most of the land is owned by the federal government. Most federal departments and agencies occupy and administer only the land occupied by and surrounding their own offices, buildings, and facilities. The two largest federal landowners are the Department of Interior and the Department of Agriculture, but the Department of Defense and the Department of Energy are also major land owners.

The federal government's three largest land managers are the Bureau of Land Management (BLM) within the Department of Interior, which oversees 270 million acres; the US Forest Service (USFS) within the Department of Agriculture, which oversees 191 million acres; and the National Park Service (NPS) within the Department of Interior, which administers 80 million acres of National Parks, National Monuments, National Recreation Areas, and other sites.

Also within the Department of Interior is the Bureau of Indian Affairs (BIA), which exists to assist Native Americans to manage their own affairs under a trust relationship with the federal government. Under various treaties, property rights on Indian Reservations generally lie with individuals or with a specific tribe or tribal entity. Thus, any inquiry regarding the search for property on Native American land should be directed to the tribal council or similar body that is associated with the specific tribe in question.

Today, the primary statutes which control the federal government's authority to deal with requests to search for treasure trove on the public lands are the Archaeological Resources Protection Act (ARPA) of 1979 (Public Law 96-95; 93 Stat. 721; 16 USC 470a et seq.) and Section 4 of the Federal Property Services Administrative Act (40 USC 310).

Since most material remains that treasure trove salvors seek to recover from the public lands are archaeological resources, ARPA is the dominant statutory authority governing their excavation and disposition. Uniform regulations found at 43 CFR Part 7 implement ARPA and define the classes of material remains subject to its provisions. Under 43 CFR Part 7.3, archaeological resources are defined as "any material remains of human life or activities which are at least 100 years of age, and which are of archaeological interest." This clearly includes remains that would, by definition, be considered treasure trove. (Black's Law Dictionary defines treasure trove as "money or coin, gold, silver, plate, or bullion found hidden in the earth or other private place, the owner thereof being unknown.")

Access to Department of Agriculture and Department of Interior land is generally open with minimal restriction, although treasure hunting and/or removal of artifacts is restricted.

Excavation of archaeological resources on BLM land without a BLM-issued cultural resource use permit is prohibited if these resources are located on public lands. Archaeological resources on public lands recovered in the course of an authorized excavation remain the property of the United States in perpetuity.

Excavation and recovery of material remains that are less than 100 years of age are protected by the Antiquities Act of 1906 (16 USC 432) and the BLM's Organic Act, the Federal Land Policy and Management Act (FLPMA; 43 USC 1701). A fifty-year threshold is a consideration that comes into play for the BLM only when determining the eligibility of a cultural property for the National Register of Historic Places [Section 101 of National Historic Preservation Act (NHPA) and 36 CFR Part 60] and deciding whether such a property is subject to the provisions of Section 106 of the National Historic Preservation Act of 1966 (NHPA; 16 USC 470), and its implementing regulations, 36 CFR Part 800. Section 106 requires federal agencies to consider the effects of their actions on historic properties, and is intended to avoid unnecessary harm to historic properties from federal or federally approved actions. Ordinarily, cultural properties that are less than fifty years of age are not eligible for the National Register, nor are they subject to the provisions of NHPA.

If archaeological resources, including treasure troves, are associated with Native American human remains or burials, the provisions of the Native American Graves Protection and Repatriation Act (NAGPRA; Public Law 101-601; 104 Statute 3048; 25 USC 3001) come into play. Since passage of NAG-PRA in November 1990, the BLM must comply with Section 3 of the act, which places ownership or control of Native American human remains and "associated funerary objects" with Native Americans. Associated funerary objects are defined in section 2(3)(A) as "objects that, as a part of the death rite or ceremony of a culture, are reasonably believed to have been placed with the individual human remains either at the time of death or later." Any treasure trove found in association with Native American burials would be subject to repatriation or disposition to a culturally affiliated Indian tribe or Native Hawaiian organization.

The second statute which controls the government's authority to deal with a potential salvor's offer to recover treasure trove from the public lands is Section 4 of the Federal Property Services Administrative Act, 40 USC 310. 40 USC 310 authorizes the Administrator of the General Services Administration (GSA) to "make such contracts and provisions as he may deem for the interest of the Government, for the preservation, sale, or collection of any property, or the proceeds thereof, which may have been wrecked, abandoned, or become derelict, being within the jurisdiction of the United States, and which ought to come to the United States, and in such contract to allow compensation to any person giving information thereof, or who shall actually preserve, collect, surrender, or pay over the same, as the Administrator of General Services may deem just and reasonable. No costs or claim shall, however, become chargeable to the United States in so obtaining, preserving, collecting, receiving, or making available property, debts, dues, or interests, which shall not be paid from such moneys as shall be realized and received from the property so collected, under each separate agreement." (Revised Statute subsection 3755; June 2, 1865, Public Law 97-30, subsection 4, 79 Statute 119.)

While the GSA has never issued regulations implementing 40 USC 310, they are responsible for issuing contracts to search for treasure trove located on federal lands. Of course, such contracts are subject to the land managing agency's approval and incorporation of any provisions or restrictions. The gross value of any treasure trove recovered, exclusive of items that are determined to be archaeological resources, is normally shared on a 50-50 percent basis with the offeror.

As 40 USC 310 states, contracts for the search and recovery of treasure trove are entered into with the stipulation that no cost and expense be incurred by the government. Potential salvors, on the other hand, may be liable for considerable expense in meeting all the legal and agency requirements. The GSA requires payment of a nonrefundable service charge of $500 to cover administrative costs of processing treasure trove requests. Potential salvors may also be required to post an appropriate bond to cover an agency's costs associated with an approved search. Also, a salvor is required to comply with Section 106 of the NHPA, which requires a salvor to obtain the services of appropriately qualified specialists to complete an archaeological inventory and evaluation of any search area to be disturbed.

Before deciding whether or not even to permit a search, GSA and the land managing agency require the submission of a current USGS or similarly scaled map showing the exact location of a search area and documentation and/or evidence that provides sufficient information to allow the land managing agency to determine the validity of the proposal. Such documentation might include maps, historical accounts, photographs, and/or any other records that would help validate a salvor's claim. A salvor must also be prepared to describe the treasure trove that he or she expects to find, estimate the value of such trove, and delineate the search methods to be employed in the recovery.

Upon receipt of the above information, GSA sends it to the jurisdictional agency for a decision on whether to allow a search. If BLM is the jurisdictional agency, it will deny or approve the search based on the assessment of the supportive documentation that a salvor provides to GSA. Particularly important is the submission of corroborative data which makes it possible for the BLM to verify with reasonable certainty that treasure trove, strictly speaking, might be found in a search area; if there is any indication that archaeological resources are the target of the salvor's search, the search will generally be denied.

The ARPA and the Federal Property Services Administrative Act apply regardless of the time that the archaeological resource and/or the treasure trove was abandoned or deposited on the public lands.

However, not all federal agencies follow identical procedures with respect to processing requests by salvors to search for treasure trove. The Fish and Wildlife Service (FWS), for example, has developed regulations at 50 CFR Part 27 prohibiting certain acts on national wildlife refuges; among the prohibited acts is the search for buried treasure, treasure trove, and valuable semiprecious rocks, stones, or mineral specimens (see 50 CFR Part 27.63) unless specifically permitted under the FWS's regulations found at 50 CFR Part 26, subpart D-Permits. The FWS, thus, is set up to prohibit the search for treasure trove under their own internal regulations unless this use is specifically authorized. Because the BLM does not have its own regulations setting out prohibited acts on the public lands, they work through GSA's regulations and appeal procedures for processing treasure trove applications.

While BLM areas are usually readily accessible, National Park Service areas are more tightly controlled and usually require an entrance fee. National Parks are areas that are set aside by Congress to "conserve the scenery and the natural and historic objects and the wildlife therein and to provide for the enjoyment of the same in such manner and by such means as will leave them unimpaired for the enjoyment of future generations." Therefore, any object or item found in an NPS area is considered to belong to all of the people of the United States, and not just the individual who found it. Consequently, the hunting for or taking of any objects in NPS areas is generally prohibited. The burying of treasure would also be prohibited in NPS areas under 36 CFR 2.1—preservation of natural, cultural, and archeological resources. In addition, the use of metal detectors is specifically prohibited by regulation which can be found at 36 CFR 2.1(a)(7).

Regulations also prohibit the possessing, removing, or disturbing of any mineral resource in all parks unless specifically permitted for a particular park area. A recreational activity along these lines is gold panning. This is currently permitted by regulation only at the Whiskeytown unit of Whiskeytown-Shasta-Trinity National Recreation Area in northern California and also in a few park areas in Alaska.

The National Forest System (NFS) lands managed by the US Forest Service, which include National Wilderness Areas as well as National Forests are generally less restrictive than National Parks. Some people have expressed concern with apparent discrepancies between the Archeological Resource Protection Act (ARPA) and US Forest Service regulations found at 36 CFR 261.2. The ARPA as amended refers to items 100 years or older. A person cannot be prosecuted under this statute if the items in question are less than 100 years old. But ARPA is not the only statute protecting historic resources.

Title 16 of the US Code, Section 551 grants the Secretary of Agriculture the authority to make provisions for the protection of NFS lands. Title 36 of the CFR, Part 60, describes agencies' responsibilities to evaluate historic resources for the National Register of Historic Places for which a property must be fifty years or older. Based upon this, 36 CFR 261.2 defines archaeological resource as "any material remains of prehistoric or historic human life or activities which are of archaeological interest and are at least fifty years of age, and the physical site, location, or context in which they are found."

This is seen as not being in conflict with ARPA, but rather an addition to it. Persons can be prosecuted under 36 CFR 261.9, which prohibits damage to or theft of property including items of historical significance that are fifty years or older.

The fifty-year time period exists only for sites (the archeological context), not general forest areas where no artifact would be expected to be found. However, people do not normally want to search in areas where there is no chance of locating anything. Therefore, the very fact of looking for artifacts is a conflict with the regulation because any item located could be within an identifiable site.

Searching for buried treasure most often includes the use of a metal detector. The use of metal detectors on NFS lands, outside of developed recreation sites, requires specific authorization. Use of metal detectors within developed recreation sites is generally allowed without authorization unless a special closure for the area is in effect. A person interested in using a metal detector, or collecting on NFS lands should contact the office having jurisdiction on the lands they plan to visit to determine the specific regulations that may apply.

Department of Defense land consists primarily of military bases. Access to such bases for any purpose is usually restricted severely, and indeed at some military areas (most of them well posted) the "use of deadly force is authorized" to deal with trespassers. Access for the purpose of digging for treasure would be virtually impossible, but a written inquiry addressed to the base commander would be a first step in determining whether an exception could be made. The name, rank, and address of the base commander (or an alternate channel) could probably be obtained by phoning the base or by inquiring at the main gate of the base.

Open land administered by the Department of Energy typically includes nuclear weapons facilities and is as restrictive, if not more, as military bases. Again, a phone call or a main gate inquiry would be a good place to start, but the odds for access are effectively nonexistent.

STATE OWNED AND/OR MANAGED LAND

As with federally owned and/or managed land, states have severely restricted or banned the excavation or appropriation of artifacts on public lands without permission. Activities involving tampering with graves, cemeteries, or human remains are illegal in all states. In some cases, permits may be obtained for paleontological researchers to excavate prehistoric remains, but these are severely restricted.

As one might imagine, the laws governing treasure hunting, while generally similar, vary from state to state in their specifics. For example, the use of metal detectors is restricted or banned in some states and widely acceptable in others.

In general, one may assume that the governments of each state to consider, as the Texas Historical Commission states, that "unregulated treasure hunting is inimical to the best interest and stated policy of the state, because of its potential for adversely affecting significant cultural resource sites.''

In order to learn as much as possible about the specific laws of the states for this book, we contacted the Office of Attorney General in all fifty states. Since this is the department charged with enforcing the laws of the state, one would naturally assume that they could tell you what the state laws state. For the most part, however, these offices responded that they were not authorized to provide legal advice to private individuals or businesses (which was not our request). Apparently their lack of authorization includes telling you what the law says. Six states were very helpful and supplied detailed information in reply to our request. These were Louisiana, Maryland, Montana, North Dakota, Texas, and Virginia. Five states, Mississippi, Nebraska, North Carolina, Oregon, and South Dakota provided limited information. We have included a synopsis of their reply in each relevant section and an address to write for further information or clarification.

We would advise you to consider it illegal (probably a felony) in all states to disturb grave sites or human remains while in search of treasure, or under any other circumstances. As noted above, unintentional discovery of human remains of any kind, whether or not they appear to be prehistoric, should be reported immediately to the local police or sheriff's office.

Most of the states urge one to seek the services of an attorney practicing law in the state and rely upon his or her advice. This is always a good idea. Another suggestion that we would offer would be to contact the sheriff's office in the county where the treasure is thought to be located.

PRIVATE LAND

Persons searching for buried treasure on private land need to have the permission of the private land owner. Obey ''No Trespassing'' signs and do not cross fences without permission. Unfortunately, in open, unfenced country, people do not realize when they cross property boundaries, but it is still your responsibility to know where you are. While property owners are generally lenient with people who wander across unfenced property lines accidentally, they are not legally obligated to condone anything that appears to be trespassing. They certainly will take a dim view of people digging on their land and removing items of value. While the courts have generally ruled that it is illegal or improper to shoot and/or kill trespassers, you should not attempt to sacrifice your life to test case law.

The origins of abandoned, lost, and mislaid property, all of which have very specific legal definitions, are rooted in English common law. Over the years, legal title to these various categories of property have been decided by the courts. In some cases, in the "absence of statute," the finder has been found to be entitled to items that "all the world save the true owner," didn't know about. When in doubt, check with local authorities and/or consult an attorney who understands the applicable local laws.

4

WHEN THE SHOVEL HITS
THE STRONG BOX

You have finally done it. Many people before you have tried and have failed, but you have done it. With a dull clank, your shovel has struck the strong box buried over a hundred years ago in a windswept coulee by a desperate, frightened thief who intended to come back for it, but who died from lead poisoning after saying too much to preserve his life but too little to allow his assassins to accomplish what you have now achieved.

You gasp, you scream, you hyperventilate, and dig with your bare hands to free this thing from its resting place. After what seems like hours, the box is free. It feels like it weighs a ton. It is agony to move such a thing, but the weight is a good sign. You try to break off the padlock with the tip of your shovel like they do in the movies, but it doesn't work. Even century-old steel will resist such an assault unless it is rusted through.

Finally, through the intercession of a tire iron and hacksaw, the padlock yields and the box comes open. The brilliance of the sun reflected from the golden surface for the first time since your grandfather's father was a young man, is blinding. You are rich, fabulously rich, or so it seems.

When the initial burst of euphoria begins to fade, you'll want to find a way to turn the treasure into spendable money. Long gone are the days when you could walk into a saloon and buy a round of drinks with a bag of gold dust, and those twenty-dollar gold pieces you've found were once common, but today they would be unacceptable as tender in most retail transactions. Most bank tellers wouldn't know what they were.

Assuming that you have read Chapter 3 and are absolutely certain that you have the right to take possession of the treasure, what do you do next? Where do you go to authenticate the value of the old coins, gold bullion, and other valuables?

Gold, silver, and copper coins are commonly and freely traded in the United States, and many have collectable value in addition to the value of the metal in them. Some dates on coins are highly sought by collectors and will occasionally be worth much more than a coin of equivalent weight having a more common date.

The first thing that you should *not* do when you have discovered gold coins is to wash or clean them. This may actually decrease their value. They should be cleaned professionally by a reputable coin dealer.

How much is the treasure worth? The spot price for gold, silver, and other precious metals changes daily and is published each day in major newspapers. This price, typically running between $400 and $500 for an ounce of gold in the 1990s, is the "melt" value of "raw" gold. On top of this, dealers will pay a "premium" depending on the collectable value of the piece. This premium is usually governed by the condition of the piece, or the date, although an "old" piece is not necessarily more valuable than a newer one. There are certain twentieth-century coins that are more valuable than many nineteenth, and some eighteenth-century coins.

As an example, common twenty-dollar gold pieces weigh 96 percent of an ounce, which was the value of gold when they were minted, and when the value of dollars was keyed to the value of gold. In the mid-1990s, their "melt" value was between $400 and $500, but depending on condition and date, they could be worth between $2,000 and $5,000 apiece. There was one case in 1995 when several were found in freshly minted condition with a rare date. They were encrusted in dirt, but were professionally cleaned, and they sold for $20,000 apiece.

Premiums are paid not only on coins, but on bullion and bars, and even on nuggets. There are many dealers and collectors who will pay a premium on nuggets if they are particularly attractive.

Where do you go to sell the treasure?

Unfortunately, banks are not typically equipped to deal with anything other than paper transactions and currently issued currency and American coins. If you have a personal relationship with a banker whom you trust, it wouldn't hurt to ask for his or her advice or referral.

Aside from this, a good place to start is to look at listings under "Coin Dealers" or "Gold Dealers" in the yellow pages of your phone book, or that of a major city near your home. The larger the city, the more choices you are likely to have.

Coin dealers also often purchase bars or bullion, or can refer you to someone who does. You should call several to get a feel for someone with whom you'd be comfortable doing business. In addition to buying coins and precious metals, such dealers will offer an appraisal service and cleaning service for a fee. Coin dealers will also be able to advise you on the value of United States currency. Antique United States currency typically has collectable value above face value, but Confederate and most foreign currency usually does not.

Some coin dealers may also deal in antiques and collectibles, including gold jewelry and precious stones. They will often be able to appraise these objects, or refer you to a reputable antique dealer who can.

In choosing a coin dealer, you should look for one who is a member of a national professional trade organization that screens the dealers and helps to maintain a high level of honesty, integrity and fairness. The following organizations can provide a list of members in your area, or can tell you whether a specific dealer is a member.

American Numismatic Association (ANA)
818 North Cascade Avenue
Colorado Springs, Colorado 80903-3279
(719) 632-2646

Numismatic Guaranty Corporation of America (NGC)
PO Box 1776
Parsippany, New Jersey 07054
(201) 984-6222

Professional Coin Grading Service (PCGS)
PO Box 9458
Newport Beach, California 92658
(714) 833-0600

Professional Numismatic Guild (PNG)
3950 Concordia Lane
Fallbrook, California 92028
(619) 728-1300

5

THE TREASURES OF NEW ENGLAND

MAINE

Maine's vivid beauty, particularly its rugged coastline, makes it a favorite of vacationers, but during the eighteenth and nineteenth centuries, its hundreds of remote islands and inaccessible inlets made it a land of opportunity for pirates looking to stash their loot, as well as people wishing to hide loot *from* the pirates. Hancock County, home to Acadia National Park, was a favorite because of its many islands.

Various eighteenth- and nineteenth-century pirates, including that notorious buccaneer, Captain William Kidd, are said to have stashed plunder at many places along the eastern seaboard, including the islands and inlets of Maine. These legends may include variations on tales of a single site which have been embellished to the point where they are now thought to be two or more treasures in the same area. Penobscot Bay, and on the lower Penobscot River area, seem to have been popular areas. One of the more intriguing involves the tantalizing "Captain Kidd's Money Cove" on Isle au Haut, which is in Knox County but is reachable by ferry from Stonington in Hancock County.

Sagadahoc County is particularly rich in both pirate and buried treasure lore. Perhaps the county's most intriguing stories involve what is known as the Portuguese Seaman's Cache,

which is rumored to include $50,000 in gold and silver coins.
They are said to be near the ruins of an old tavern on the north
end of St. John's Island in Casco Bay.

Hancock County

- Baron Castine's Treasure: Located on Swan's Island in the
  Kennebec River.
- The Buried Treasure of Codlead Marsh: Said to be at or near
  Codlead Marsh, in the Penobscot Bay area.
- Various Buried Pirates' Treasures: In and around the mouth
  of the Penobscot River near Bucksport, off US 1.
- The Lost Indian Gold Mine: In the general area of Lead
  Mountain Ponds and Lead Mountain.
- Captain Kidd's Treasure: In the vicinity of Musselridge
  Channel in Penobscot Bay, and on the lower Penobscot
  River area.

Kennebec County

- Captain Kidd's Treasure: In the vicinity of Hallowell and
  near Pittston.

Knox County

- Captain Kidd's Money Cove: On Isle au Haut (reachable by
  ferry from Stonington in Hancock County).
- Buried Pirate Treasure: On Monroe Island, on the west side
  of Penobscot Bay.

Lincoln County

- Captain Kidd's Treasure: Located on the Sheepscot River,
  near Wiscasset.
- Samuel Trask's Buried Treasure: Located near or in Edge-
  comb.

Sagadahoc County

- Captain Kidd's Treasure: At Boothbay Harbor, Cliff Island, and at Jewell's Island.
- Captain Kieff's Treasure: Located on Cliff Island.
- Dixie Bull's Treasure: Located on Cushing Island.
- Edward Lowe's Treasure: Located on Pond Island in Casco Bay.
- Various unnamed Pirate Treasure: Located on Bailey Island in Casco Bay, on Great Chebeague Island, on Haskell Island off Harpswell Neck, and on Outer Hebron Island.
- The Portuguese Seaman's Cache: Located near the ruins of a tavern on the north end of St. John's Island in Casco Bay.

Somerset County

- Buried Pirate Treasure: In the vicinity of Skowehegan Falls, on the Kennebec River, near the fork of US 201 and 201A.

Waldo County

- Captain Kidd's Treasure: In the vicinity of Frankfort.
- Timothy Barrett's Treasure: At or near Liberty.

Washington County

- Pirate Bellamy's Treasure: In the Machias Bay area.

York County

- Blackbeard's Treasure: Smuttynose Island (*see also* New Hampshire).

MASSACHUSETTS

Massachusetts was the largest and most populous of the New England colonies, and Boston was a major metropolis and trading center by the beginning of the eighteenth century.

With this accumulation of wealth, there was plunder, and with this came the need to hide the loot. Much of it remains hidden.

According to legend, there are major concentrations of treasure in the counties around Boston, including the Atlantic coastline of Essex and Norfolk Counties. The legendary Captain Kidd figures prominently in stories of pirate treasure, although such sites attributed to him may be the plunder of other buccaneers, and credited to him because his name is more notorious in the literal sense of the word. Other pirates are also credited, of course. Captain John Quelch buried a cache of gold and silver coins on Snake Island, off Cape Ann, in the Isle of Shoals chain north of Boston in Essex County.

Another region rich in treasure lore is Berkshire County in western Massachusetts. While pirate caches are the rule on the coasts, the Berkshires are said to contain stashes of loot looted by the British—as well as the Patriots—during the Revolutionary War. British soldiers spent several months raiding settlements in the northwestern corner of the state after their defeat at Saratoga in 1777. With the Patriots closing in, they buried three wagon loads of plunder off the road that is now Route 9, near Dalton in Berkshire County. Meanwhile, about ten miles north of Dalton, several chests of gold and silver coins were taken from a Hessian supply train by a band of Patriots and buried on Mt. Amos, northeast of Cheshire.

A well known and poignant tale unfolds with the massacre of the people of Stockbridge in Essex County. During the French and Indian War, the town was raided and burned to the ground but, before they were killed, the residents buried their valuables near the town. Many were never recovered.

Another tantalizing tale of the Revolutionary War era involves $175,000 (now worth many times that) in British gold sovereigns that was buried under a rock marked with an ''A'' in or near Byfield, along the banks of the Parker River, west of what is now the Blue Star Memorial Highway.

Barnstable County (Cape Cod)

• The Buried Treasure of Money Hill: Located in Provincetown.

- Pirate Treasure: Said to be located at or near Oyster Harbor.
- The Chatham Treasure Chest: In the vicinity of Chatham.

Berkshire County

- British Plunder: In the vicinity of Dalton, off Route 9.
- The Black Grocery Gang Loot: Located on Mt. Washington, southeast of Pittsfield.
- Mohegan Indian Treasure: At or near Bash-Bish Falls, near Mt. Washington.
- Captain Kidd's Treasure: In the vicinity of Cheshire.
- The Hermit's Buried Treasure: Located on Greylock Mountain.
- The Pirate Treasure Chest: Buried on the banks of the Hoosic River near Cheshire on Route 8.
- The Sand Mill Buried Treasure: Said to be hidden near Dalton.
- The Hessian Payroll: Located on Mt. Amos, northeast of Cheshire.
- Stockbridge's Caches: Buried near Stockbridge along what is now Route 102.

Around Boston Bay

- Captain Avery's Chest of Diamonds and Gold Coins: Said to have been buried on Gallops Island.
- Captain Billie's Lost Treasure: Located on Grape Island.
- Captain Kidd's Treasure: Buried on Conant's Island.
- John Breed's Buried Treasure: Said to have been buried on Swan Island.
- John Quelch's Pirate Treasure: Said to be buried on Snake Island.
- The Money Bluff Buried Treasure(s): Hidden on Deer Island.
- William Marsh's Buried Treasure: Located somewhere on Apple Island.

- The Nantasket Beach Treasure: Located on the south shore of Boston Harbor.
- Pirate Treasures: Stories tell of various buried or concealed caches located on Calf Island, on Castle Island, on Grape Island, on Great Brewster Island, on Little Brewster Island, and on Hog Island.

Bristol County

- The Watson Pond Cache: Gold coins and jewelry hidden near Watson Pond, on the outskirts of Taunton, near Route 44.

Essex County

- The Byfield Pirate Treasure: In the vicinity of Byfield.
- Harry Main's Buried Treasure: Located on Plum Island.
- The Gorrill Brothers' Cache (aka the Danny Fry's Hill Treasure): In or near Tenney Castle ruins, on Danny Frye's Hill in Methuen.
- The Isle of Shoals Treasure: Captain John Quelch's cache on Snake Island, off Cape Ann, in the Isle of Shoals chain (*see also* New Hampshire).
- The Parker River Cache: Under a rock marked with an "A" in or near Byfield, west of the Blue Star Memorial Highway along the banks of the Parker River.
- Thomas Veale's Hoard: Several chests hidden near Dungeon Rock Cave, at the mouth of the Saugus River, near Lynn.

Franklin County

- Captain Kidd's Gold and Jewels: A chest of gold and jewels buried by the swash-buckling captain near Turner Falls, on the Connecticut River, off Route 2.

Hampshire County

- Alden Culver's Buried Treasure: Said to be hidden in or near West Chesterfield.

- The Buried Gold Bullion: In a cave near South Hadley off State Route 5.

Middlesex County

- The Brink's Robbery Loot: In the vicinity of Somerville.
- Devil's Den Buried Treasure: In the vicinity of Wilmington.
- Spanish Treasure: At Harbor Pond, east of Townsend, off State 119.
- The Old Fort Treasure: $400,000 in gold and silver coins hidden near Shirley, Route 111, Middlesex County.
- The Willard Tavern Hoard: $100,000 in gold and silver coins buried near the old Willard Tavern in Shirley.
- Thomas Smith's Farm: Buried on the farm, on the east bank of the Assabet River off Route 62 near Maynard.

Naushon Island

- The Buried Treasure of Tarpaulin Cove: Located on Naushon Island, north of Martha's Vineyard.

Plymouth County

- The Brockton Caches: Several caches of gold coins were buried around a rich shoe manufacturers' mansion in Brockton, off Route 24, east of Stonehill College.

Suffolk County

- The Buried Treasure of Shirley Point: Located on Shiley Point south of Winthrop.
- Willard's Taproom Treasure: Said to have been located in Winthrop, Willard's Taproom may or may not be confused with the old Willard Tavern in Shirley.

NEW HAMPSHIRE

A largely rural state with a short coastline, New Hampshire is dominated by the rugged White Mountains in the north that blend into rolling hills in the south toward the state's major population centers.

One of the areas most important legends recalls that while fleeing the American Revolution in 1775, English Governor John Wentworth buried a strongbox filled with gold and silver coins along with six chests of silver and gold plate in a wooded area between Portsmouth and Smithtown in Rockingham County.

Other stories swirl around the rugged Isles of Shoals that New Hampshire shares with Maine and Massachusetts. Numerous treasures have been buried here, both by the crews of wrecked ships and by pirates. Blackbeard buried caches on Londoner and Smuttynose islands, while John Quelch buried gold and silver on the west side of Appledore Island, as well as on both Snake and Star islands. Captain Sandy Gordon buried plunder on White Island.

Carroll County

- The Lost Mine of Ossippee Range: In the vicinity of Whittier.

Grafton County

- The Treasure of the Worcester Cave: In the vicinity of Campton.

Hillsborough County

- Captain Kidd's Buried Treasure: At Rye Pond near Antrim.
- John Cromwell's Cache: Buried at his trading post on the west side of the Merrimack River near Merrimack, two miles above Cromwell Falls off US 3.

The Isles of Shoals

- Various Caches: Blackbeard buried loot on Londoner and Smuttynose islands; John Quelch buried gold and silver on the west side of Appledore Island, as well as Snake and Star islands; and Captain Sandy Gordon buried plunder on White Island (*see also* Maine and Massachusetts).

Orange County

- John L. Wood's Cache: Buried near his sawmill, near Woodsville, at the junction of the Connecticut and Ammonoosuc rivers.

Rockingham County

- The Hemp Spinner's Buried Treasure: In the vicinity of Portsmouth.
- Governor John Wentworth's Hoard: In an area between Portsmouth and Smithtown that would have been wooded in 1775.

Strafford County

- John Clifton's Pirate Booty: Buried near Durham off Route 108 and close to the Oyster River.

VERMONT

The only New England state without an Atlantic coastline, Vermont is not without legends of pirate treasure, including one attributed to the great Captain Kidd. Nevertheless, the Green Mountain State is filled with other tales, especially involving her granite crags and deep canyons, which verge on inaccessible because they are choked with vegetation in summer and with snow and ice in winter. Indeed, like several western states, rugged little Vermont boasts not one, but *two* "Hell's Half Acres."

Like western Massachusetts to its immediate south, much of Vermont's treasure lore dates back to large batches of coins and such stolen by one side or the other during the Revolutionary War. A group of Americans took $90,000 in gold and silver coins from a British supply train west of Bennington, and several kegs of silver coins were buried by the British on

Harmon Hill, five miles southeast of Bennington.

In 1773, before the war, $75,000 in gold coins were buried near Cedar Beach during an attack by Native Americans and allegedly never recovered. During the French and Indian War, Native Americans being pursued by the British hid several cart loads of stolen booty on Sable Mountain.

Nearly a century later, during the Civil War, daring Confederate raiders robbed three banks in St. Albans of $114,522 and buried the loot near Highgate Springs, just below the Canadian border.

Addison County

- The Buried Treasure of Hell's Half Acre: In the vicinity of Bristol.

Bennington County

- The Patriot Raiders Treasure: $90,000 in gold and silver coins buried west of Bennington off US 7 near the Bennington Battle Monument.
- The Harmon Hill Silver: Kegs of silver coins buried by the British on Harmon Hill, 5 miles southeast of Bennington.

Chittenden County

- The Lost British Treasure: $75,000 in gold coins buried off US 7 near Cedar Beach.

Essex County

- The St. Francis Treasure: $40,000 in "church plate" and gold and silver coins, allegedly hidden along the west bank of the Connecticut River off Route 105 near Bloomfield.
- The Spanish Treasure of Winooski Valley: Stashed somewhere in Winooski Valley.

Franklin County

- The St. Albans Bank Robbery Loot: $114,522 buried near Highgate Springs, just off US 7 near Lake Champlain.

Lamoille County

- The Lost Slayton Gold Mine: Somewhere on Mount Mansfield in the Green Mountains.

Orleans County

- The Hell's Half Acre Treasure: $375,000 in British gold and silver coins, buried on the south end of Lake Memphremagog off US 5 near Newport.
- The Lost Treasure of Providence Island: Providence Island, Lake Memphremagog.

Rutland County

- The Lost Birch Hill Silver Mine: Between Sherburne Pass and Pico Peak, about 15 miles northeast of Rutland on US 7 County.
- The Money Cave: Rumored pirate's treasure is buried in or near Money Cave in the Green Mountains off US 7 east of South Wallingsford.
- The Sable Mountain Booty: Several cart loads hidden on Sable Mountain, 10 miles north of Sherburne Pass, which is in Windsor County. (Note: There is also a Sable Mountain in northern Essex County.)

Washington County

- The Lost Treasure of Camel's Hump: In the vicinity of Waterbury, probably near a hill that reminded someone of a camel's hump(s).

Windsor County

- Levi Bailey's Mill Cache: Located near his mill, on a stream between Mount Ascutney and Ready off Route 106.

Winham County

- The Buried Pirate's Treasure: In the vicinity of Whitingham.
- Captain Kidd's Buried Treasure: In the vicinity of Bellows Falls.
- The Pagan Treasure of St. Francis: Located near White Hills.

RHODE ISLAND

The islands and waterways in and around Narragansett Bay and Rhode Island Sound were traditionally great favorites of eighteenth-century pirates. Indeed, nearly every island and inlet has a story involving Captain William Kidd. Several of the more important sites are included below.

Newport County

- The Joe Bradish Treasure: Said to be on Block Island.
- Captain Kidd's Treasure: On Block Island.
- Captain Kidd's Treasure: Buried in Pirate's Cave on Conanicut.
- Captain Kidd's Treasure: Somewhere on Hog Island.
- Captain Kidd's Treasure: Near Sakonnet at the mouth of the Sakonnet River.
- Captain Kidd's Breton Point Cache: Located on Brenton Point in Rhode Island Sound.
- Charles Harris' Chest: Buried along the beach at the base of Newport Cliffs, south of Newport.
- Thomas Tew's Cache: $100,000 in booty, buried somewhere in Newport.

Washington County

- Captain Kidd's Treasure: In the vicinity of Watch Hill.

CONNECTICUT

Like its neighboring states, Connecticut has a treasure lore that is deeply rooted in the eighteenth century with legends of both the American Revolution and the pirates who once called upon these shores in the dead of night with mist scudding across the face of a full moon. As with other New England states, notably Maine and Rhode Island, the name of Captain William Kidd is featured prominently in the lore.

Perhaps the most interesting legend of loot in the legacy of the Constitution State involves a cache of Revolutionary War coins. Supposedly $2 million in gold coins were specially minted for George Washington's army and buried during a Tory and Indian attack near a tavern that was once owned by Captain Lemuel Bates, just one mile north of East Granby off what is now Route 202.

Fairfield County

- Captain Kidd's Island Treasure: Money Island, Pilot Island, and Sheffield Island, off Norwalk.
- Captain Kidd's Buried Treasure: Stratford Point near Milford.

Hartford County

- Captain Kidd's Treasure: In the vicinity of Wethersfield.
- Tyron's Landing Treasure: Tyron's Landing near Wethersfield.
- The Lost Revolutionary War Coins: One mile north of East Granby off Route 202, near the site of a tavern once owned by Captain Lemuel Bates.

Middlesex County

- Captain Kidd's Cache: In the vicinity of Middletown.
- Captain Kidd's Treasure: Said to be buried at Kelsey Point.

New Haven County

- Captain Kidd's Buried Treasure: In the vicinity of Milford.

New London County

- Captain Kidd's Hidden Cache: In the vicinity of Old Lyme.

6

THE TREASURES OF THE NORTHEAST

New York

Today, as it has been for the better part of a century, the Empire State's largest city is the financial capital of the world. But if you look north from the upper floors of the lower Manhattan skyscrapers where today's billion-dollar deals are made, you can see on the distant horizon the beginning of the rolling hills and the woodlands of upstate New York that have witnessed many an event that has spawned the legends of American folklore. Washington Irving spun his tales here. Was the "headless horseman" real? Were Brom Bones or Rip van Winkle based on real characters? Most of the legends are just that, but many legends also have a basis in fact.

What then about the legends of treasure, accumulated and hidden in the same spirit of strictest confidence as the deals cut each day on Wall Street? What gold and silver lies today in quiet hiding places in the upstate woods?

As the stories go, much was hidden, lost and/or forgotten here during the Revolutionary War. A defecting British paymaster hid $150,000 in gold coins near the northern tip of Lake Colton and there are at least three treasures near or around Ticonderoga, which itself is a site of legendary importance in American history. These three include a cache of gold and silver coins buried near old Fort Amherst at Crown

Point about twelve miles north of Ticonderoga, a British cache near the banks of Long Pond, and Native American plunder buried a few miles north of Ticonderoga in the Mount Pharaoh Range.

John and Walter Butler were a pair of Tory raiders—outlaws actually—who robbed from American patriots during the war and buried their loot somewhere near their homestead, the Butlersbury Mansion, which is on Switzer Hill near Fonda in the Mohawk Valley of Montgomery County. Many treasures were stolen, hidden, and/or lost when both the British and the Patriots destroyed most of the buildings in the area around Saratoga Springs.

During the French and Indian War, which preceded our Revolution by two decades, French raiders supposedly buried fifteen chests of treasure on Grand Island in the Niagara River between Buffalo and Niagara Falls. The story of the "Mad Frenchman" originated during the same period. He was actually a merchant named Clairieux (or Clairvieux) who is said to have buried several kegs of coins near the ruins of his house and store, also on Grand Island in the Niagara River. He may have been mad and he may have been crazy, but he had the gold, and he may not have recovered it.

There are also several important legends of treasure hidden during the eighteenth century on islands in the St. Lawrence River. Before he surrendered to the British during the French and Indian War, the French commander of the garrison on Isle Royal buried his treasure somewhere on the island, possibly opposite Waddington. Downstream near Louisville, Britain's Lord Amherst reportedly buried $100,000 on tiny, but aptly named, Treasure Island.

And then there are the stories of troves that have been lost, not through the actions of thieves or raiders, but through natural occurrences. An example is the Lost Blenheim Silver Mine. It was sealed during the Revolutionary War, but a subsequent landslide completely hid the mine's location on the north side of Mount Utsayanth in Schoharie County.

Some legends of lost treasure are long shots as far as the treasure is concerned, but are worth mentioning as tales. One such story is the Sulphur Springs Treasure. At the turn of the twentieth century, a travelling "medicine man" buried two chests near the Sulphur Springs Health Resort between North

Pitcher and Pitcher in Chenago County. Whether the chests contained patent medicine, snake oil, or something of greater value, the story does not tell. There may, however, be someone around the county today who could shed more light on the subject.

The "Roaring Twenties" also gave rise to their share of tales. Few gangsters made the twenties and thirties roar louder than the notorious Dutch Schultz (his real name was Arthur Flegenheimer) who controlled the bootleg booze trade in New York City and much of the United States from his base in the Bronx. Targeted by the cops, rival thugs and G-men alike, Schultz was finally gunned down in Newark's Palace Chop House by Charlie "The Bug" Workman in October 1935. But during the height of his career, Dutch had stashed his "retirement fund" in a grove of pine trees on the banks of the Escopus Creek in the Catskill Mountains, five miles south of Phoenicia in Ulster County. The "fund" reportedly was a large iron box with over $7 million in gold coins (worth over $70 million, and possibly as much as a billion today), negotiable bonds, currency, uncut diamonds, and jewelry.

Great fortunes were made in bootleg booze because of Prohibition. Some of the money was invested as bootleggers turned "legitimate," but some was stashed and lost. Such was the case of $2.5 million in currency (silver certificates, valued in 1920 dollars) hidden on an abandoned farm on the north end of Ashokan Reservoir, a mile east of Shokan in Ulster County.

Legends of treasure hidden by the famous and infamous are always colorful. One such treasure is that which is said to have been hidden by the great automobile magnate Walter Percy Chrysler before his death in 1940, on or near his estate at Southampton Beach on Long Island.

More recently hidden is Moses Follensby's hoard, which includes $400,000 in gold coins and currency which he buried on his property near Follensby's Road, two miles southeast of Tupper Lake in Franklin County. Tracy Maxwell, who died in 1948, left a will stating that $135,000 in currency (in 1948 dollars) and the family jewelry was buried near his barn, two miles west of Surprise in Greene County. At about the same time, a wealthy recluse is said to have hidden $750,000 on his estate in Hicksville, Long Island.

Albany County

- Outlaw Gold: In or near Wynd Cave, near Knox, off Route 157A.

Chenago County

- The Sulphur Springs Treasure: Two chests were hidden near the Sulphur Springs Health Resort, between North Pitcher and Pitcher off Route 26.

Erie County

- The Lost British Payroll (aka The British Paymaster's Lost Treasure): Gold coins hidden near Mount Colden near the town of Colden.
- The Lost Indian Gold Mine: In the vicinity of Mount Colden.

Essex County

- The Fort Amherst Treasure: Buried near old Fort Amherst at Crown Point.
- The British Cache: Located near the banks of Long Pond.
- The Native American Plunder: Buried a few miles north of Ticonderoga in the Mount Pharaoh Range.

Franklin County

- Moses Follensby's Hoard: Gold coins and currency hidden on his property near Follensby's Road, 2 miles southeast of Tupper Lake, near Routes 3 and 30.

Greene County

- Captain Kidd's Sleepy Hollow Treasure: In the vicinity of Palenville.
- Tracy Maxwell's Treasure: Currency and the family jewelry

was buried near his barn, 2 miles west of Surprise, off Route 81.

Hamilton County

- The Adolphus Lavigne Lost Mine: Said to be somewhere in the county.

Montgomery County

- John and Walter Butler's Plunder: Located near Butlersbury Mansion, on Switzer Hill near Fonda in the Mohawk Valley.

Nassau County, Long Island

- The Hicksville Hidden Treasure: Located on an estate in Hicksville near Route 135.

Niagara County

- French and Indian War Treasure: Located on Grand Island in the Niagara River between Buffalo and Niagara Falls.
- The Mad Frenchman's Treasure: Located on Grand Island in the Niagara River between Buffalo and Niagara Falls.

Orange and Putnam Counties

- Captain Kidd's Crow's Nest Treasure: In the vicinity of the US Military Academy at West Point.
- Captain Kidd's Treasure: At Kidd's Point on the Hudson River. (The story may refer to a point on the Hudson in Putnam County or Westchester County.)

Rockland County

- The Letterrock Mountain Treasure: In the vicinity of Bear Mountain.
- Captain Kidd's Grassy Point Treasure: Located near Stony Point.

St. Lawrence County

- The Isle Royal Treasure: Somewhere on Isle Royal in the St. Lawrence River off the port of Waddington County.

- Treasure Island: In the St. Lawrence River between Louisville and Rooseveltown near Route 37.

Saratoga County

- British Plunder: Located on the north shores of Lake Saratoga.

- General Burgoyne's Treasure: In the vicinity of Saratoga Springs.

- Saratoga Springs's Revolutionary War Treasures: Various small treasures possibly located in or on the site of 1770s ruins in the area of Saratoga Springs near US 87.

Schoharie County

- Buried Pirate's Treasure: In the vicinity of South Gilboa.

- Native American Treasure: Hidden in a cave near Conesville, off Route 30.

- The Lost Blenheim Silver Mine: One of several mines on the north side of Mount Utsayantha, it was sealed off during the Revolutionary War with a subsequent landslide completely obscuring the mine's location.

- Native American Silver Mine: Reportedly on Blenheim Mountain, it may be the same as the Lost Blenheim Silver Mine.

- The Lost Schoharie County Silver Mines: Other mines on the north side of Mount Utsayantha.

Seneca County

- The Loomis Outlaw Gang Loot: $40,000 buried in Montezuma Swamp near Seneca, off Route 96.

Suffolk County, Long Island

- Charles Gibbs's Hoard: At Southampton Beach near Route 27 and the end of Long Island.

- Walter Chrysler's Cache: Located on his estate, just above Southampton Beach.

- Joe Bradish's Cache: The stories indicate its being stashed at the "tip" of Long Island, but are unclear as to whether this means near Montauk Point or Orient Point.

Sullivan County

- The Lost Shawangunk Silver Mine: Between Lake Mongaup and Hodge Pond in the Catskill Mountains.

- Tory Raiders' Plunder: In a cave in the Shawangunk Mountains near Summitville, off Route 209.

Ulster County

- The Bootleg Cash: $2.5 million in currency hidden on an abandoned farm on the north end of Ashokan Reservoir, a mile east of Shokan, Route 28.

- Dutch Schultz's Retirement Fund: A large iron box with over $7 million in gold coins, negotiable bonds, currency, uncut diamonds, and jewelry buried in a grove of pine trees, on the banks of the Esopus Creek, in the Catskill Mountains, 5 miles south of Phoenicia near Route 214.

- The Lost Truman Hurd Gold Mine: Located near Shokan.

- Rufe Evans' Cache: Silver bullion and ore buried in or near Accord on Route 209.

- The Schlechtenhorst Lost Silver Mine: Within 5 miles of Woodstock off Route 212.

- The Tongorara Treasure: Over a million dollars in gold and silver coins hidden between the Tongorara Reservoir and Kingston near US 87.

Warren County

- The Lost Nippleton Silver Mine: Somewhere in Warren County.

Washington County

- Robert Gordon's Fortune: The fleeing Tory buried $75,000 in the marshes called The Haven, on the Poultney River, a mile northeast of Whitehall near what is now Route 4.

Westchester County

- The Kidd's Point Pirate Treasures: At Kidd's Point on the Hudson River, in a cave overlooking the Hudson River near Stoney Point, south of Peekskill (*see also* Orange and Putnam Counties).
- Money Hill: Treasure buried at "Money Hill," in or near Croton-on-the-Hudson.

Wyoming County

- A Depression-Era Hoard: $45,000 is said to have been buried on the Wilbur Rogers farm, two miles west of Warsaw near Route 20A.

NEW JERSEY

Like that of its neighbors, New Jersey's treasure lore is a mix of pirate caches and loot stolen by—or hidden from—raiders in conflicts dating back to the eighteenth century. For example, retreating British troops buried two chests of gold coins on Apple Pie Hill, three miles southwest of Chadsworth in Burlington County, and plunder taken by the British during the sacking of Philadelphia was hidden near Palmyra in Camden County. In 1778, over $5,000 in gold and silver coins was captured from the British and buried by the Patriots somewhere near Hopewell in Mercer County.

Pirates' treasure is rumored to be buried all along New Jersey's coastline. Several caches, including one credited to Captain William Kidd, are buried at Cliffwood Beach on Raritan Bay, and there are other rumored "Captain Kidd" treasures at Sandy Hook and Red Bank in Monmouth County. The legendary Joe Bradish, who was also active along the coast of Long Island, is reported to have had several caches on the Jersey shore.

Twentieth-century troves include that of Vincent Conklin, who died in 1921 after burying his treasure near the barn on his farm a mile east of Tabernacle on Route 532 in Burlington County. Also during the 1920s, bootleggers buried about $1.5 million near Lake Manetta, south of Lakewood near Route 9.

Atlantic County

- John Bacon's Abescon Cache: At the northern end of Abescon Island facing Reeds Bay.

Burlington County

- Blackbeard's Treasure: In or near the town of Burlington.
- Conklin's Cache: Located near the barn on the farm owned by Conklin in 1921, 1 mile east of Tabernacle on Route 532.
- Revolutionary War Gold: Two chests of gold coins buried on Apple Pie Hill, 3 miles southwest of Chadsworth near Route 563.

Camden County

- British Plunder: Hidden near Palmyra and Route 453.

Cape May County

- Giles Shelly's Buried Treasure: In or near the town of Cape May.
- Joe Bradish's Treasure: Located on or near Turtle Gut Inlet.
- Captain Kidd's Lilly Pond Cache: At or near Lilly Pond near Cape May Point.

Cumberland County

- Joe Bradish's Treasure: Located on the Cedar River, 1 mile southwest of Cedarville near Route 553.

Essex County

- The Murdered German Officer's Treasure: In the vicinity of Caldwell.

Hudson County

- Hendrick Dempster's Hoard: $50,000 in gold coins buried on the farm formerly owned by Dempster off US 95 near North Bergen.

Mercer County

- The Hopewell Treasure: Over $5,000 in gold and silver coins buried in 1778 by Patriots somewhere near Hopewell on Route 518.

Middlesex County

- Captain Kidd's Cache: Located on Cliffwood Beach at Raritan Bay near Route 35.

Monmouth County

- Captain Kidd's Monmouth Treasures: Located on Sandy Hook and Red Bank.
- The Treasure of the Pine Robbers: Somewhere in Monmouth County.

Ocean County

- John Bacon's Treasure: In the vicinity of Barnegat Lighthouse, at the north end of Island Beach, 10 miles north of Ship Bottom.

- The Prohibition Money: $1.5 million buried near Lake Manetta, south of Lakewood near Route 9.
- The Treasure of the Barnegat Pirates: Located on Island Beach.

Salem County

- James Gilam's Treasure: Located on the farm he once owned at Finn's Point on the Alloway River, near Quinton on Route 581.
- The Seven Star Tavern: Several caches are buried in or near the site of the Seven Star Tavern on the north bank of the Salem River near Sharptown.

Sussex County

- The Bunker Kid's Buried Loot: Located near Newton on Route 206.
- Arthur Barry's Depression Hoard: $100,000 in gold coins and currency buried on the farm near Andover on Route 517 that he owned in the 1930s.
- The Hanover Neck Cache: $50,000 in gold coins is said to have been buried in the mid-nineteenth century under a big tree on a farm near Hanover Neck, 6 miles east of Morristown on Route 10, which was owned by a Henry Walker in the mid-twentieth century.
- William Besthorn's Estate: $12,000 in gold coins buried on the estate near Caldwell on Route 506.

PENNSYLVANIA

The treasure lore of the Keystone State includes tales that date from before the Revolutionary War to the middle of the twentieth century. Among the earliest is a legend of buried Spanish treasure at Spanish Hill on the Susquehanna River about a mile south of Sayre in Bradford County. During the American Revolution, a British spy confessed to burying a cache of gold coins near a tavern on Pond Street in Bristol across from Burnington Island in the Delaware River. In 1775, a paymaster buried $150,000 in gold and silver coins on Laurel Hill, close to Fort Necessity, near Farmington in Fayette County. In 1812, during the Anglo-American rematch, Colonel Noah Parker stashed a reported $5 million in gold and silver bullion along the banks of the Potato River, near Crosby in McKean County.

Stories of eighteenth-century pirate loot and treasure salvaged from sinking ships is a favorite staple of the genre along eastern seaboard coasts and waterways, and one of the more colorful tales is that of the "War of Jenkin's Ear" Treasure. As the story goes, twenty-two tons of mercury in ceramic flasks and 38,000 pieces of eight taken from the merchantman *San Ignacio El Grande* and were buried on the west bank of the Delaware River, two to three miles southwest of Chester. A pirate story is told in the legend of Blackbeard's Treasure. This is not *the* famous pirate, but rather the British Admiralty's Captain Blackbeard, who may well have been a pirate in his own way, but who is said to have buried $1.5 million in silver bars near Gardeau in McKean County.

In 1863, during the Civil War, Gettysburg, Pennsylvania was the scene of the great battle that blunted the Confederate drive into the North, but at the same time that Robert E. Lee was mounting his massive offensive, small groups of Confederate troops were active elsewhere in Pennsylvania. This involved not only disrupting military communications, but a bit of banditry as well. Often the raiders netted more in their raids than they could carry home and they had to stash it in the North. For the most part, the raiders never again set foot in Pennsylvania and the loot was never recovered. One such Civil War trove consists of plunder packed in barrels and buried near Mountain House, close to the top of the Snowshoe Mountains in the Alleghenies, about one mile west of Wingate on Route 220. Confederate guerrillas also hid fifteen tons of silver bullion in a sealed cave two miles north of Uniontown in Fayette County.

In McKean County and thereabouts, they still talk of the 1890 Emporium Bank Robbery, after which $60,000 in gold coins and currency was buried beside the Kinzua railroad bridge, crossing the Kinzua Creek, five miles northeast of Mount Jewett. As far as we know, it was never recovered. The same is true of outlaw David Lewis's Lost Saddlebags, which were not actually lost off his horse, but buried with $10,000 in gold coins on the bank of the Juanita River at Juanita Terrace in Mifflin County. Lost but not forgotten, they were probably carried away by high water.

The story of a 1948 Mount Carmel Plane Crash is still fresh in the minds of a few folks around Northumberland County. According to one of the survivors, just before the crash a passenger threw $25,000 out of the window. It probably landed on Mount Carmel near Ashland.

Berks County

- The Cache of Captain Doanes' Raiders: $100,000 in plunder buried in a well, 1 mile south of Wernersville, Route 422.

Bradford County

- Buried Spanish Treasure: at Spanish Hill on the Susquehanna River, 1 mile south of Sayre near Route 220.

Bucks County

- The British Spy's Cache: In Bristol, near the present or former site of a tavern on Pond Street, across from Burnington Island on the Delaware River.
- Doctor John Bowman's Buried Plunder: Located on Bowman's Hill, along the Delaware River in the Washington Crossing State Park.

Centre County

- Outlaw David Lewis's Cache: Described as being buried within sight of the jail in Bellfonte on Route 220.
- The Mountain House Confederate Treasure: Buried in barrels near Mountain House, close to the top of Snowshoe Mountains in Allegheny Range, 1 mile west of Wingate near Route 220.

Delaware County

- The War of Jenkin's Ear Treasure: twenty-two tons of mercury in ceramic flasks and 38,000 pieces of eight buried on the west bank of the Delaware River, 2 to 3 miles southwest of Chester.

Fayette County

- The Lost Payroll on Laurel Hill: $150,000 in gold and silver coins on Laurel Hill, close to Fort Necessity and Farmington on Route 40.
- The Confederate Raiders' Silver: fifteen tons of silver bullion hidden in a sealed cave two miles north of Uniontown near Route 40.

Fulton County

- Lost Indian Mines: Located near McConnellsburg in the Cove and Dickeys Mountains.

Luzerne County

- Michael Rizzalo's Loot: $12,000 buried near Laurel Run Creek in the Laurel Run Mountains, four miles east of Wilkes-Barre.

McKean County

- The Other Blackbeard's Treasure: $1.5 million in silver bars buried near Gardeau on Route 155.
- The 1890 Emporium Bank Robbery: $60,000 in gold coins and currency buried beside the Kinzua railroad bridge, crossing the Kinzua Creek 5 miles northeast of Mount Jewett near Route 6.
- Colonel Noah Parker's Treasure: $5 million in gold and silver bullion, buried in 1812 along the banks of the Potato River near Crosby on Route 46.

Mifflin County

- Joe Fracker's Cache: Gold and silver coins buried along the banks of the Jacks River, north of Lewistown on Route 22.
- Outlaw David Lewis's Lost Saddlebags: $10,000 in gold coins buried on the bank of the Juanita River at Juanita Terrace south of Lewistown (probably since carried downstream by high water).

Northumberland County

- Raiders' Loot: Buried in a cave overlooking the Delaware River, 2 miles north of Easton off Route 22.
- The 1948 Mount Carmel Plane Crash: $25,000 thrown from an airplane window over Mount Carmel near Ashland.
- The Lost Bear Jasper Mine: Somewhere along the Delaware River near Raubville and Route 32.

Tioga County

- The Tory Raiders' Treasure Chests: Several chests are buried on the McMillan Farm near the Tioga River, 2 miles east of Bloomsberg on Route 15.

DELAWARE

With a long, sandy, and relatively secluded coastline, Delaware has always been a favorite of pirates and buccaneers of every stripe. Both Captain William Kidd and the notorious James Gillian, a member of Kidd's crew, supposedly buried treasure on Kelley Island in Delaware Bay off what is now the Bombay Hook National Wildlife Refuge. In about 1700, another pirate of legend, the feared Blackbeard (his real name was Edward Teach) is said to have buried treasure on the banks of Blackbird Creek, near the present site of the town of Blackbird on Route 10. The vicious and dreaded slave trader, Patty Cannon, also hid ill-gotten gains on Delaware shores.

The "New Castle Pirate Treasures" are several supposed sites in the county of the same name where pirates hid their treasure. These sites include a deep well near the former town jail and another in the area of Taylor's Bridge in the town of New Castle.

Those who despair that lost treasures are never found will take heart in the news that one of the Harrington Hoards— several caches of Spanish silver coins buried near Harrington— was uncovered by a bulldozer operator in 1973.

Kent County

- James Gillian's Cache: Located on Kelley Island in Delaware Bay off the Bombay Hook National Wildlife Refuge.
- Captain Kidd's Treasure: In the vicinity of James Gillian's Cache off on Kelley Island near the Bombay Hook National Wildlife Refuge.
- The Harrington Hoards: Several hoards of a rich merchant buried near Harrington on Route 13.

New Castle County

- The New Castle Pirate Treasures: Several sites in and around the town of New Castle including one in a deep well near the town jail and another in the area of Taylor's Bridge.
- Blackbeard's Treasures: Located on the banks of Blackbird Creek, near Blackbird on Route 10.

Sussex County

- The Patty Cannon Treasure: About $100,000 in gold and silver coins buried near Laurel on Route 24.
- Silver Hill: A wealthy planter buried a large hoard of silver bullion and coins a few miles south of Milford.

DISTRICT OF COLUMBIA

The Commandant's Treasure: About $25,000 in gold and silver coins are alleged to be buried under the Commandant of Marine Corps former home near 8th and ''I'' streets in Washington, DC.

MARYLAND

Maryland's geography rambles from the wrinkled hills and valleys of the Appalachians to the sandy Atlantic shore, and its treasure lore ranges from pirate plunder to loot looted by both sides in the Civil War. The latter era is represented in caches said to be located on the grounds of the Croissant Mansion (aka the Resurrection Mansion) in St. Mary's County and on the Resurrection Plantation in Somerset County.

In addition to the stashes of pirates, slave traders also used the Maryland shore to bury loot. The hated and sadistic slaver Patty Cannon hid $100,000 in gold coins in several caches southeast of Federalsburg, on the Harold Smith farm in Reliance, and possibly near an old tavern on the Nanticoke River near Riverton in Wicomico County.

Of the fifty states that we contacted for clarification of laws governing treasure hunting, Maryland was one of eleven that supplied useful information. In Maryland, most statutes for the protection of cultural resources on state owned or controlled lands are administered by the Maryland Historical Trust, which is the State Historic Preservation Office. The requirements and protections vary for terrestrial (solid ground) and submerged lands. It is the policy of the State Forest and Park Service to safeguard the archeological resources under its care.

In regard to private property, other than one possible exception and the laws of trespass, collecting and excavating on private property is allowed by the owner or with the owner's permission. The exception is outlined in Article 83B, Section 5-621. This allows private property owners to petition the state for cultural resource protections, like those applicable to state lands, to apply to their property.

It is not illegal to own or to use a metal detector on State Forest and Park Service public swimming beaches within established parameters. These include obtaining permission from the park or forest manager, using them between 9 am and dusk from Labor Day through May 30. During the summer months parks often permit use before 9 am and after dusk but this varies from park to park and should be checked first. Parks reserve the right to limit their use at other times at the discretion of the park manager (for example during a special event). It is not illegal to use metal detectors or to otherwise collect on private property with the permission of the landowner. It is not illegal to use metal detectors in state owned or controlled waters. (This includes all tidal waters and virtually all nontidal waters. When in doubt check first.)

However, regardless of whether a metal detector is involved or not, it is illegal on state-owned or controlled land or waters to excavate, destroy, or substantively injure an historic property or its environment; endanger other persons or property; or violate any other applicable regulations or provisions of state, federal or local law.

It is important to understand that the Submerged Archaeological Historic Property Act applies to resources (sites, features, objects, etc.) that are 1) submerged, 2) archaeological, and 3) historic. It does not apply to modern coins, jewelry, watches, keys, or any other resource that is less than 100 years old and not eligible for the National Register of Historic Places.

Permits are required for activities involving excavation or disturbance of submerged archaeological historic properties. The requirements for obtaining a permit vary according to the type of archaeological historic property and the proposed activities.

The Submerged Archaeological Historic Property Act of 1988 is implemented by regulations that came into effect January 2, 1993. The Act and Regulations allow for public access to "inspect, study, explore, photograph, measure, record, conduct a reconnaissance survey, or otherwise use and enjoy a submerged archaeological historic property without being required to obtain a permit." However, as noted above, it is not permitted to excavate, destroy, or substantively injure an historic property or its environment, endanger other persons or property, or violate any other regulations or provisions of state, federal, or local law.

It is also permissible, without a permit, to collect from any one site up to five individual artifacts that 1) are exposed or resting on the bottom sediments of submerged lands, 2) do not require excavation to recover, and 3) do not cumulatively weigh more than twenty-five pounds. Recovery of these items is permissible only by hand or through the use of screwdrivers (maximum length, twelve inches), wrenches and pliers (maximum length, twelve inches and maximum jaw width, two inches). This may only be done once at any given site, one cannot return to the site each day for another five objects. However, a person may not collect any artifacts from a site that is designated or is eligible for designation as a National Historic Landmark or is listed or is eligible for inclusion on the National Register of Historic Places. A list of these sites is maintained by the trust and may be viewed on request. Contact the Office of Archaeology for the most current information.

In essence, it is perfectly acceptable in Maryland for hobbyists to look for and collect modern jewelry, modern coins, and so forth from nonhistoric sites. When materials that obviously are not modern are recovered, the individual may collect up to five items by hand or with the small hand tools mentioned above, then must stop and report these finds to the trust. The trust can assist in the identification of the items and determine if they are 100 years old or National Register eligible. The Trust has two weeks to assess the significance of the site, determine whether the property is indeed 100 years old or National Register eligible, and inform the finder whether (s)he must obtain a permit to continue work there, or whether the trust will request the area be avoided, or whether the area is not likely to yield significant historic information and the public may continue to collect there.

If the site is significant, the person will be informed that further work on the site requires a permit, or that the site has been determined as one of those on the state's register of sites for which permits will not be issued. At present, there are no submerged sites for which the state will not consider a permit application. The trust can also advise on recommended conservation methods to ensure the survival of the objects recovered.

When in doubt, or for more information, contact

The Office of Archaeology
Division of Historical and Cultural Programs
100 Community Place
Crownsville, Maryland 21032
(410) 514-7661

Baltimore County

- Jean Champlaigne's Treasure: Located near Catonsville.
- The Druid Hill Park Treasure (aka Captain Kidd's Cache):

In Druid Hill Park in the city of Baltimore.

- The Old Mansion Home's Hidden Cache: $65,000 in gold coins at the Old Mansion Home in the city of Baltimore.
- Jacques Champlaine's Hoard: Champlaine buried a chest containing $150,000 in gold coins on his farm near Catonsville, close to the Old Frederick Road.

Caroline County

- The Poor House Treasure: $30,000 in gold coins hidden in the ghost town of Plaindealing, 2 miles east of Hurtock near Route 331.
- Patty Cannon's Caches: $100,000 in gold coins in several caches southeast of Federalsburg on the Dorchester County line, and on the Harold Smith farm near Reliance on the border with Sussex County, Delaware.

Carroll County

- The Lost Indian Silver Mine: Located on Rattlesnake Hill near Silver Run on Route 140.

Dorchester County

- Jake Hole's Treasure: Hole buried $200,000 on the south bank of the Choptank River, near Lodgecliffe 3 miles northwest of Cambridge.

Frederick County

- The Monocacy River Treasure: $100,000 in gold coins buried about 2 miles south of Frederick, on the banks of the Monocacy River near Route 144.
- Old Hagan's Tavern Treasure: Buried chest containing $38,000 in gold coins and jewels near old Hagan's Tavern, midway between Braddock and Braddock Heights on Route 40.

Prince George's County

- The Treasure in the Well: A safe containing the money from a bank robbery in the 1960s was placed in a well on a farm near Upper Marlboro. (Rumors that the safe was recovered have not been verified).

St. Mary's County

- The Croissant Mansion Treasure: Somewhere on the grounds of the mansion of the same name in California on near Route 235.

Somerset County

- The Resurrection Plantation Treasure: Hidden in or near the plantation house, which itself is near Kingston and Route 413.

Talbot County

- The Goldsborough Creek Treasure: Located on the eastern shore of the Miles River.
- The Old Friends Meeting House Treasure: $50,000 in gold coins buried near the Old Friends Meeting House, 1 mile east of Easton.
- The Tred Avon Treasure: Located on the eastern shore of the Tred Avon River.

Wicomico County

- Patty Cannon's Tavern Treasures: Located on the site of an old tavern on the Nanticoke River near Riverton.

Worcester County

- Cellar House Pirates' Caches: Various pirates are said to have buried their loot at or near Cellar House on what is now US 113, midway between Berlin and Snow Hill.

7

THE TREASURES OF THE SOUTH

VIRGINIA

As the home of England's first American colony, the home of four of the first five American presidents and the capital of the Confederacy, Virginia has had a rich and turbulent history. Through it all, there were fortunes made and fortunes lost. There were also fortunes hidden, and many of them also became lost. The origin of these fortunes was not always in commerce. There were working gold and silver mines in Virginia, and these mines still exist. Native Americans worked several gold mines on Mount Rogers in Grayson County, as well as several silver mines near Passage Creek Gap in Shenandoah County. Native Americans also hid raw gold and plunder in and around the Shenandoah Caverns, five miles north of New Market in Shenandoah County.

A major part of the Old Dominion's treasure lore is involved with great fortunes in Confederate gold that was hidden and subsequently lost when the state was being mauled as a major battleground during the Civil War.

Because of the destruction of the cities such as Roanoke, Lynchburg, Richmond, and Petersburg during the war, many citizens never recovered their personal caches. While much of this has been unearthed over the years, much still remains to be found. A rumored $4 million in gold bullion and coins was hidden by a Confederate general and some slaves somewhere on the McIntosh Farm, west of Lynchburg. Iron chests containing 2,921 pounds of gold, 5,088 pounds of silver, and $13,000 in gems and jewelry were hidden near Montvale, northeast of Roanoke. The gold alone would now be worth over $18 million.

Some stories are more specific. Captain John Mosbey, a Confederate guerrilla, buried a treasure worth millions in gold, silver—from coins to church plate—and weapons between two large pine trees, between Culpeper and Norman. Other Civil War–era treasures are supposed to be buried on the grounds of the Carter's Grove Plantation, six miles southeast of Williamsburg.

Perhaps the most intriguing Civil War story involves the case of $10 million in gold coins and bullion which the British government loaned to the Confederacy. It was buried along the banks of the James River, near the Berkeley Plantation east of Hopewell in Prince George County, and never reclaimed. If subsequent flood waters ate into the banks, the coins could have been swept downstream and/or buried under many yards of silt, although the bullion would probably be at the original location.

With the Confederacy long since gone, one wonders who would be responsible for the interest on this loan, and whether the United States government would recognize any claim on the treasure by the British government since the Confederacy was a national entity never recognized by the United States.

Of special importance for the lightly industrialized South during the Civil War were its saltpeter mines, which yielded an essential component of gunpowder. One of these mines had a further function. Abraham Smith buried $60,000 in an abandoned saltpeter mine during the Civil War. Located in the Poor Valley, between Allison's Gap and Saltville near Route 9, the mine is not known to have given up its treasure.

One of Virginia's most interesting pirate treasures is one for which an actual set of directions exists. In the manuscript, which dates from the early eighteenth century, the buccaneer Charles Wilson wrote of the $2 million treasure hidden on Assateague Island, "Ye treasure lies hidden in a clump of trees near 3 creeks lying 100 paces or more north of the 2nd inlet above Chincoteague Island.

The directions seem simple until one ponders the map of Assateague Island, now managed by the National Park Service as the Assateague Island National Seashore. North of Chincoteague Island, which fits into Assateague like a ball into a socket, there are dozens of larger inlets, hundreds of smaller inlets, and an uncountable number of often-seasonal streams.

Pirates also worked out of Hog Island Bay, twenty miles north of Cape Charles, and there have also been numerous shipwrecks on the shores of Hog Island.

A more recent treasure involves the theft from the US Bureau of Printing & Engraving in 1911 of $31,700 in new currency and a set of twenty-dollar printing plates. Packed in a leather pouch, it was hidden near a creek, north of Warrenton in Fauquier County.

Of the fifty states that we contacted for clarification of laws governing treasure hunting, Virginia was one of eleven that supplied useful information. The office of Virginia's Assistant Attorney General in the Real Estate and Construction Section of the Office of the Attorney General states that there is no statutory authority for that office to give legal advice to private individuals in matters such as treasure hunting on public lands, but offers the suggestion that treasure hunters review the Virginia Antiquities Act (Code § 10.1-2300 et seq.), the Estrays and Drift Property Act (Code § 55-202 et seq.), and the Uniform Disposition of Unclaimed Property Act (Code § 55-210.1 et seq.).

It is important to note that entering upon, or disturbing, any state-owned land may be subject to rules and regulations of the agency which controls that property. If there is any uncertainty as to which agency controls a particular parcel of land, contact

Office of the Director
Department of General Services
202 North Ninth Street, Suite 209
Richmond, Virginia 23219

Accomack County

- The Pirate Charles Wilson Cache: $2 million on Assateague Island (Assateague Island National Seashore) in a clump of trees near three creeks lying 100 paces or more north of the second inlet above Chincoteague Island.

Bedford County

- The McIntosh Farm Treasure: $4 million in gold bullion and coins hidden somewhere on the McIntosh Farm, west of Lynchburg, 1 mile south of Forest, near Route 811.
- The Beale Treasure: Iron chests containing 2,921 pounds of gold, 5,088 pounds of silver, and $13,000 in gems and jewelry, hidden near Montvale, 4 miles north of the Buford Tavern ruins, 12 miles northeast of Roanoke.

Culpeper County

- Captain John Mosbey's Cache: Gold, silver (from coins to church plate), and weapons between two large pine trees, between Culpeper and Norman, close to Route 522.

Fauquier County

- The Bureau of Engraving Treasure: $31,700 in new currency and a set of twenty-dollar printing plates in a leather pouch, hidden near a creek north of Warrenton, off Route 211.

Gloucester County

- Carter's Grove Plantation: Treasures buried on the grounds of Carter's Grove Plantation 6 miles southeast of Williamsburg on US 60.

Grayson County

- The Mount Rogers Lost Mines: Native American gold mines on Mount Rogers.

Henrico County

- Buried Confederate Treasure: Said to be hidden many places in and around Richmond.

Louisa County

- The Boswell's Tavern Treasure: Revolutionary War–era treasure is buried somewhere on the tavern property, located on the South Anna River near Boswell Village.

Northampton County

- Hog Island Bay: Possible pirate treasures and numerous shipwrecks on the shores of Hog Island.

Prince George County

- The Lost British Loan Money: $10 million in gold coins and bullion loaned to the Confederacy was buried along the banks of the James River, near the Berkeley Plantation, east of Hopewell near Route 10.

Prince William County

- William Kirk's Hoard: The pirate buried $50,000 in silver coins on the Snow Hill Farm, 1 mile south of Bristow near Route 619.

Rockingham County

- The Peaked Mountain Buried Treasure: Located on a "peaked mountain" near McGaheysville.

Shenandoah County

- The Lost Indian Silver Mines: In the vicinity of Passage Creek Gap and Strasburg on US 81.
- Shenandoah Caverns: Native Americans hid raw gold and plunder in and around the Shenandoah Caverns 5 miles north of New Market on Route 11.

Smythe County

- The Saltpeter Mine Treasure: $60,000 buried in an abandoned saltpeter mine in the Poor Valley between Allison's Gap and Saltville near Route 9.

Warren County

- Powell's Fort: Edwin Powell buried treasure near the natural rock fortress called Powell's Fort, under a rock with a horseshoe carved on it, in the Fort Mountains near Waterlick in the Shenandoah Valley.

WEST VIRGINIA

The western counties of Virginia broke from the "Old Dominion" in 1861 and joined the Union as a separate state in 1863 even as the Civil War raged through the gaps and "hollers" of the new state. Known as the "Mountain State" because of its rugged terrain, West Virginia proved a good hiding place from its earliest moments as an independent state. It was so good that many were never found.

In 1763, during Pontiac's War, while the area was still part of British Virginia, a farmer buried his gold and silver in the "old burying ground" near Charleston.

Going back further, Native Americans worked gold and silver mines in the area of Raleigh and Mason Counties as well as elsewhere. Before abandoning their diggings, they are believed to have sealed silver bullion and other valuables inside a mine between Workman's and Meadow creeks, near Beckley in Raleigh County.

Grover Berdoll stashed his caches in the Potomac Valley. During the Civil War he buried $150,000 in gold and silver coins, along the west bank of the Potomac near Harper's Ferry in Jefferson and Berkeley Counties. Another early 1860s trove involves $300,000 in Confederate gold bullion buried near Bear Camp Run, along the left fork of the Buckhannon River, one mile down river from Palace Valley in Upshur County.

Events in the twentieth century have also added to the Mountain State's treasure lore. In 1928, bank robbers buried $50,000 on the campus of the Davis & Elkins College, south of Elkins near Route 290. A few years later in the 1930s, $2,000 in gold and silver coins was buried on the Carpenter farm, a half mile from Bear Fork Creek near Groves Hollow. It was also during the 1930s that Kelly Cooper buried his life's savings near his barn. The Cooper farm is on the east bank of the Tug Fork River in Mingo County. Also on the east bank of the Tug Fork, which forms the border with Kentucky, is $200,000 in gold coins buried by Dennis Atkins near the toll bridge north of Kermit.

Berkeley County

• A Grover Bergdoll Treasure: In the Potomac River Valley.

Gilmer County

• The Carpenter Farm Cache: $2,000 in gold and silver coins buried on the Carpenter farm, a half mile from Bear Fork Creek, near Groves Hollow.

Jefferson County

• A Grover Berdoll Cache: $150,000 in gold and silver coins buried on the west bank of the Potomac River near Harpers Ferry on US 340.

Kanawha County

• The Old Burying Ground Treasure: Gold and silver buried in 1763 at the "old burying ground," Charleston.

Logan County

• The Farmer's Fortune: Several iron chests buried along the Guyandot River, one half mile north of Chapmanville on Route 10.

Mason County

- The Lost John Smith Silver Mine: East of Point Pleasant near Route 62.

Mingo County

- Kelly Cooper's Cache: Located near the barn on the Cooper farm, on the east bank of the Tug Fork River, near a place called Sprigg on Route 49.
- Dennis Atkins's Hoard: $200,000 in gold coins buried near the toll bridge, on the east bank of the Tug Fork River, north of Kermit on US 52.

Raleigh County

- The Indians' Lost Silver Mine: Silver bullion and other valuables inside a sealed mine between Workman's and Meadow Creeks near Beckley.

Randolph County

- The Lost Bank Loot: $50,000 in bank loot buried in 1928 on the campus of the Davis & Elkins College, south of Elkins.

Upshur County

- A Confederate Gold Cache: $300,000 in Confederate gold bullion buried near Bear Camp Run on the left fork of the Buckhannon River, 1 mile down river from Palace Valley.

NORTH CAROLINA

The treasure lore of the Tarheel State includes mentions of both Civil War–era loot in the hills and pirate plunder along the shore, though North Carolina also has gold and silver mines. There are, for example, numerous caches hidden near the old mining cabins on Richardson Creek, a branch of the

Rocky River, near the Anson Mine, five miles south of Wadesboro. There are also "Lost Indian Mines"—such as the Lost Sontechee Indian Silver Mine—between Marble and Andrews in the Snowbird Mountains. There is the so-called Cherokee Treasure, $100,000 worth of gold and silver bullion buried in a sealed cave in the Great Smoky Mountains, three miles west of Smokemont in Swain County. Cherokees are also said to have buried twenty mule loads of silver bullion within Nantahala Gorge on the Nantahala River near Lauada in Graham County.

Eighteenth-century pirates crossed the shores and history books of every state on the eastern seaboard, and North Carolina is no exception. Indeed, the area around Elizabeth City was the main base of operations in the United States for buccaneer Edward Teach, better known as Blackbeard, who probably cached loot throughout the area. One possible "Blackbeard" site is the old colonial ruins on the south bank of the Pasquotank River, three miles north of Elizabeth City, but there are many rumored locations for valuables cached by Teach in other counties, several of which are detailed below.

During the era when Teach was active, there were hoards he probably wished he'd had a crack at. In one such instance, survivors from the wreck of a Spanish galleon buried 200,000 pesos in gold and silver at Cape Lookout, fifteen miles southeast of Morehead City in Carteret County. In 1781, twenty chests and kegs of plunder were buried by British raiders on the banks of Abbot Creek, north of the US 64 bridge, east of Lexington Davidson County and in 1854 William D. Wentworth died in his sleep after burying his satchel of gold coins in the woods behind the Brummel's Inn on the old stage road between High Point and Greensboro in Guilford County.

A tantalizingly specific tale tells of fifty pounds of emeralds in a bag that are said to be hidden between two rocks near the Caler Fork bridge, ten miles north of Franklin. In another popular legend, the $500,000 embezzled from the Milwaukee Bank is hidden in or near a fishing lodge on the Cape Fear River, south of Tarheel.

Tales of hidden Civil War gold includes cooking pots filled with gold coins that are buried along the old Southern Railway line between McLeansville and Burlington. A reported total of $100,000 in gold coins are said to be stashed in several places in Charlotte. One cache is near where Sugar Creek passes under Moorehead Street and another is rumored to have been hidden under an old house at Elizabeth Street and Hawthorne Lane. Two chests of Confederate silver and cash are buried along the Deep River, two miles south of Ranseur near Route 64 in Randolph County.

Even though today's public thrives on tabloid journalism and scandals in high places, few people remember the Teapot Dome Affair, much less the notorious sex scandals that swirled around President Warren G. Harding in the 1920s. Mrs. Harding, a paranoid, but a paranoid with cause, hired former Justice Department investigator Gaston B. Means to "shadow" her husband and report on his torrid peccadillos with young Nan Britton. Harding died mysteriously—the whispers said "murder"—in August 1923 and Means was arrested two months later, allegedly as a "fall guy" for corruption higher up within the Harding administration. Means served time and his full story was lost to history. Also lost was an estimated $500,000 buried on his former estate in Concord, near Charlotte.

Of the fifty states that we contacted for clarification of laws governing treasure hunting, North Carolina was one of eleven that supplied useful information. The office of their attorney general replied that the use of metal detectors and the removal of artifacts from state parks is prohibited. For further information, refer to North Carolina General Statutes, Chapter 116B (abandoned property) or Article 3, Chapter 121 (abandoned shipwrecks).

Alamance County

- Hidden Civil War Gold: Cooking pots filled with gold coins, buried along the old Southern Railway line between McLeansville and Burlington.

Anson County

- Richardson Creek: Numerous caches are said to be hidden near the old mining cabins on Richardson Creek, a branch of the Rocky River, near the Anson Mine 5 miles south of Wadesboro off Route 109.

Beauford County

- Blackbeard's Treasure: At Plum Point, on Pamlico Sound near Bath.

Bladen County

- The Milwaukee Bank Money: $500,000 in embezzled funds hidden in or near a fishing lodge on the Cape Fear River south of Tarheel on Route 131.

Brunswick County

- ''King Roger'' Moore's Treasure: More than $375,000 is thought to be buried at the Orton Plantation on the Cape Fear River near Brunswick Town.
- The Gander Hill Plantation: $30,000 in gold and silver coins buried on the Gander Hill Plantation, west of Wilmington near Route 74.
- Pirate's Treasure: Located near Fort Caswell at the mouth of Cape Fear River.

Buncombee County

- The Summit Treasures: Several Dutch farmers' caches are hidden on Summit Mountain, 2 miles north of Black Mountain near Route 70.

Cabarras County

- Julius Benjamin's Treasure: Gold coins and jewelry hidden on Julius Benjamin's farm, 1 mile north of Mount Pleasant near Route 49.
- The Gaston B. Means Estate: $500,000 on his estate in Concord near Route 70, 14 miles northeast of Charlotte.

Carteret County

- The Cape Lookout Cache: Gold and silver from a Spanish galleon buried at Cape Lookout, 15 miles southeast of Morehead City.
- The Hidden Silver: A chest of silver coins buried among the sand dunes on Shaklesford Beach, Harker's Island.

Chatham County

- The Devil's Stamping Ground Treasure: At the "Devil's Stamping Ground," somewhere in Chatham County.
- The Williamson Plantation Treasure: Gold coins and silver plate buried at the Williamson Plantation, on the west side of the Haw River, 1 mile northeast of Bynum on Route 15.

Cherokee County

- The Lost Indian Silver Mine: Between Marble and Andrews in the Snowbird Mountains.
- The Lost Sontechee Indian Silver Mine: In the Snowbird Mountains near Andrews on Route 129. (Its entrance was sealed and therefore hidden.)

Chowan County

- Old Man Batt's Buried Treasure: In the vicinity of Edenton.

Cumberland County

- The McKinnon Plantation Hoard: $200,000 hidden on the McKinnon Plantation, 2 miles west of Fayetteville on Route 401.

Davidson County

- The 1781 British Plunder: Twenty chests and kegs of plunder buried on the banks of Abbot Creek, north of the US 64 bridge, east of Lexington.

Franklin County

- The Lost Emeralds: 50 pounds of emeralds in a bag, hidden between two rocks near the Caler Fork bridge, 10 miles north of Franklin.

Graham County

- The Nantahala Gorge Caches: Gold coins hidden near the mouth of the Nantahala Gorge and 20 mule loads of silver bullion buried within the gorge. Nantahala Gorge is on the Nantahala River, near Lauada, on Route 19.

Guilford County

- The Brummel's Inn Gold: William D. Wentworth's satchel of gold coins, buried in the woods behind the Brummel's Inn, on the old stage road between High Point and Greensboro.

Hyde County

- Blackbeard's Pirate Treasure: At Ocracoke Inlet, on Ocracoke.

Mecklenburg County

- Civil War Treasure: $100,000 in gold coins buried near where Sugar Creek passes under Moorehead Street in Charlotte.
- Civil War Treasure: A cache of gold coins was hidden under an old house at Elizabeth Street and Hawthorne Lane in Charlotte.

Moore County

- The Keener Estate: Gold and silver coins were buried during the Depression at the Keener Estate, outside of Pinehurst near Route 2.

New Hanover County

- The Gander Hall Treasure: In the vicinity of Wilmington.
- The Pirate's Lost Hoard: Located on Money Hill, between Wrightsville Beach and Kirkland near Route 17.
- Pirate's Treasure: At Buccaneer Point, on the Cape Fear River, 4 miles above Wilmington.
- Stede Bonnett's Pirate Treasure: Somewhere in New Hanover County, almost certainly on the coast.

Pasquotank County

- Blackbeard's Treasures: At the old colonial ruins on the south bank of the Pasquotank River, 3 miles north of Elizabeth City near Route 17.

Randolph County

- Lost Confederate Treasure: Two chests of Confederate silver and cash are said to be buried along the Deep River, 2 miles south of Ranseur near Route 64.
- David Fanning's Deep River Cache: His plunder is hidden in a cave along the Deep River near Guilford.
- The Campgrounds Cache: A Revolutionary War cache of gold and silver coins hidden at the Campgrounds, outside Red Crossing near Route 64.

Rutherford County

- The Lost Copper Hill Silver Mine: In the vicinity of Silver Creek Knob.

Swain County

- Cherokee Treasure: $100,000 in gold and silver bullion sealed in a cave in the Great Smoky Mountains, 3 miles west of Smokemont near Route 73.

Wake County

• Lost Indian Gold Mines: In the vicinity of Raleigh.

SOUTH CAROLINA

As with North Carolina, the treasure lore of the Palmetto State includes both pirate and Civil War–era loot. Middleton Place Gardens, the Santee River area, in or adjacent to the Cape Romain National Wildlife Refuge, is particularly rich in possible troves.

The Hampton Plantation manor house, a mile west of the Santee River Bridge near Jamestown, is said to have buried treasure, and Confederate raiders buried a stolen $100,000 Union Army payroll along the Santee River near St. Stephan. In 1781, two generations and several wars earlier, Tory raiders stashed gold and silver plunder somewhere on North Island in Winyah Bay.

Berkeley County

• Middleton Place Gardens: Treasure is said to be buried at the Gardens, on the Ashley River, 14 miles southwest of Charleston.

• The Mulberry Plantation Treasure: Located on the Cooper River, 30 miles above Charleston.

• The Hampton Plantation: Loot buried in or near the manor house, 1 mile west of the Santee River Bridge on Route 41 near Jamestown.

• The Lost Union Payroll: $100,000 buried along the Santee River, near St. Stephan near Route 45.

Charleston County

• The Drayton Hall Treasure: Located on the Ashley River near the junction of Routes 57 and 61.

Georgetown County

- The 1781 Tory Treasure: Gold and silver plunder buried somewhere on North Island in Winyah Bay.

Horry County

- The Pirate Treasure of Fort Randall: Located on Little River.

Spartanburg County

- The Walnut Grove Plantation Treasure: In a filled-in well on the plantation, located near I-26 and US 221, near the Tyger River, south of Spartanburg.

York County

- The Williamson's Plantation Treasure: Four miles east of McConnells on Route 322.

GEORGIA

Beneath the pines of the Georgia mountains lie numerous caches of gold bars, iron boxes filled with Spanish and Confederate coins, and more than a ton of silver ore buried by a band of French explorers from New Orleans in the 1760s. The reason they were exploring what is now Georgia with that much silver is lost to the mists of time.

A prominent chapter in the treasure lore of the Peach State is that of Cherokee gold. There is said to be $400,000 in gold coins buried near the ruins of a log cabin on the Chattahoochee River near the old McIntosh Trail, five miles southwest of Whitesburg. Other troves to which is ascribed a Cherokee connection are 1,000 bars of gold in a cave on Rocky Face Mountain near Varnell and caches in caves on Fishtrap Mountain, near Rock Creek Lake and in White Bubbling Springs, located in the foothills of Oak Mountain, twelve miles north of Canton.

The Cherokees are also reputed to have thrown two hundred

pounds of raw gold into the Etowah River and also to have buried a hoard somewhere around the Shallow Rock Creek Bridge near Canton. A further eight pots of "Cherokee" gold was said to have been buried on the old F. R. Grover farm between Silver City and Frogtown, west of Route 19. One of these was actually found in 1932, leaving seven still buried. For this, we'd recommend studying local newspaper accounts from 1932 for details.

This is not the only tale from Georgia that includes treasure found as well as treasure lost. The US Army's Fort Benning is located on land that included what used to be the farm of a man named Whitney. According to the legend, Mr. Whitney buried two chests filled with about $150,000 in gold on his farm. During the 1920s, one of these was unearthed from a ravine near what was then known as Outpost #1.

Georgia also has its share of pirate gold, including a Blackbeard stash or two, caches stashed by despicable slave traders. For example, $20,000 in gold coins was buried by a scarred old slaver at the Malachi homestead, a mile northeast of Subligna in the Chattahoochee National Forest.

Large sums of questionable or mysterious origin are also part of the literature A certain Mr. Ledbetter buried $40,000 in gold coins from the Dahlonega Mint in his basement in 1851. He was killed when his house and mill burned down and, as the story goes, the gold was not recovered. In another story, more than $100,000 in gold coins in a mail pouch are thought to have been hidden in a hollow near a spring on the Cob County side of the Chattahoochee River near the old Marietta Road, three hundred yards northeast of the old Nashville, Chattanooga, and St. Louis Railroad tracks. This railroad merged with the Louisville & Nashville in 1957, which itself was acquired by the Atlantic Coast Line, which was fully absorbed by CSX Corporation in 1982. Much of the trackage of bygone subsidiaries has been abandoned and pulled up by the newer railroad, but local sources may remember where the old lines ran.

Dionysus Chenault buried $200,000 in gold near the horse pen on his plantation, but died before he revealed the exact location of the money. The best we can tell you is that it is said to be twelve miles northeast of Washington near Route 44. Inquire locally about the Dionysus Chenault Plantation.

As with the other states of the former confederacy, Georgia's hills still hold treasures that date to the Civil War. For example, in 1864 during the sacking of Atlanta, some Union soldiers made off with close to $6 million in gold, silver, jewelry, and other valuables. One would imagine that this would not have been contrary to the rules of engagement laid out by General William Tecumseh Sherman who was in command of the brutal, scorched-earth "March to the Sea." Apparently, however, the actions of these soldiers were not sanctioned by Sherman and they were arrested. They were hanged without revealing where they had hidden the loot, but it is believed to be somewhere in the vicinity of Lithonia, about ten miles east of Atlanta, on what is now US 20.

Meanwhile, to prevent it from being taken by Union forces, about $3 million in gold was hidden by a man named Guy Rivers and several slaves as Sherman was approaching Dahlonega in 1864. Rivers died of a heart attack before the gold was recovered. The gold was hidden either in Guy Rivers Cave on the Josephine Mine property or somewhere along the Rocky Etowah River Bluff in Lumkin County. Gold coins and family silver were also buried during the Civil War on the Brock homesite, four miles northeast of Clermont, close to Route 129 between Gainesville and Cleveland near the Chattahoochee National Forest.

The biggest Civil War cache of all is said to be located in or near Washington in Wilkes County. Confederate President Jefferson Davis met with his cabinet for the last time in this town in 1865, and most of the $10 million in the Confederate Treasury is said to have been buried in different locations in the surrounding area. Some of the paper currency, which is of little or no value, has been found, but the gold bullion is still missing.

Baldwin County

- The Treasure of Peter William's House: In or near Milledgeville.

Bartow County

- The Felton Farm: $50,000 in gold and silver coins in a chest were buried near the barn on the Felton Farm, on Route 41 near Cartersville.

Brooks County

- The French Silver: In the 1760s, French explorers from New Orleans hid about 2,700 pounds of silver ore in the vicinity of the forks of Okapilco Creek and Mule Creek.

Bryan County

- Blackbeard's Pirate Treasure: Located on Ossabaw Island, 15 miles north of the Blackbeard Island National Wildlife Refuge.

Carroll County

- Hidden Treasure of the Cherokee: $400,000 in gold coins buried near the ruins of a log cabin on the Chattahoochee River near the old McIntosh Trail, 5 miles southwest of Whitesburg on Route 27A.

Chatham County

- The Flint House Pirate Treasure: Hidden in or near the Flint House in Savannah.

Cherokee County

- Lost Cherokee Gold: Several iron chests of gold bullion in White Bubbling Springs, in the foothills of Oak Mountain, 12 miles north of Canton.
- The Cherokees' Shallow Rock Creek Treasures: 200 pounds of raw gold thrown into the Etowah River and also a buried

hoard near the Shallow Rock Creek Bridge, near Canton.

- Ledbetter's Dahlonega Mint Gold: $40,000 in Dahlonega Mint gold coins hidden in 1851 in the basement of Ledbetter's house (since burned) near Free Home on Route 20.

Cob County

- The Mail Pouch Gold: More than $100,000 in gold coins in a mail pouch was hidden in a hollow close to a spring on the Cob County side of the Chattahoochee River, near Marietta and the old Marietta Road, 300 yards northeast of the old Nashville, Chattanooga, and St. Louis Railroad tracks.

DeKalb County

- The Union Soldier Loot: Close to $6 million in gold, silver, jewelry, and other valuables, believed to be somewhere in the vicinity of Lithonia, about 10 miles east of Atlanta on US 20.

Floyd County

- The Lindale Mills Gold: $600,000 in gold bullion is hidden in a grove of trees on Lindale Mills property west of Route 411 south of Rome.

Forsyth County

- Cherokee Gold: Eight pots of gold (one was found in 1932) on the old F. R. Grover farm between Silver City and Frogtown, west of Route 19.

Gilmer County

- Cherokee Treasure: Located near the mouth of Flat Creek on the Cossawatte River.
- Hidden Cache of the Cherokee: Located near the Scarecorn and Talking Rock Creeks.
- The Lost Indian Silver Mine: Somewhere in the vicinity of the mouth of Flat Creek near the Coosawattee River.

Gwinnett County

- The Lost Billy Chambles Gold Mine: Three miles north of Suwanee Creek near Beaver Run Creek, off Route 23.

Habersham County

- The Lost Indian Gold Mine and the Lost Hunter Gold Mine: Located near Unicoi Gap.

Lumkin County

- Guy Rivers Hidden Gold: About $3 million in gold hidden either in Guy Rivers Cave, on the Josephine Mine property or somewhere along the Rocky Etowah River Bluff.

McDuffie County

- Jeremiah Griffin's Hidden Gold: Over $100,000 in gold hidden in the banks of a creek 2 miles south of the Little River.

McIntosh County

- Blackbeard's Buried Treasure: Located on Blackbeard Island (Blackbeard Island National Wildlife Refuge/Blackbeard Island Wilderness Area) in Sapelo Sound.

Meriwether County

- The Layfield Plantation Cache: Gold and silver coins hidden under a pile of rocks near the summit of Pine Mountain, on the Layfield Plantation, 2 miles south of Warm Springs, off Route 85W.

Murray County

- The Coosawattee River Gold: Gold bars are hidden in a cave under a waterfall on the Coosawattee River east of the bridge near Fort Coosawattee, near Carter's Corner, near Route 441.

Muscogee and Chattahoochee Counties

- The Fort Benning/Whitney Farm Treasure: The old Whitney farm was on the grounds of what is now Fort Benning, and buried here were two chests filled with about $150,000 in gold, one of which was found in a ravine near Outpost #1 in the 1920s.

Rabun County

- The Cherokee Cache on Fishtrap Mountain: In the caves of Fishtrap Mountain, 5 miles northeast of Lakemont.
- Cole Rogers' Treasure: In the vicinity of Tallulah Falls in Haversham and/or Rabun counties.

Troup County

- The Lipscombe Plantation Cache: $100,000 in gold and silver coins buried on the old Lipscombe Plantation, between the west bank of the Chattahoochee River and the Alabama state line, 8 miles north of La Grange.

Union County

- The Cherokee Lost Cache: Located near or in Rock Creek Lake, between Suches and Gaddiston.

Walker County

- The Slave Trader Gold: $20,000 in gold coins buried by a slave trader at the Malachi homestead 1 mile northeast of Subligna (in Chattooga County) on Route 201 in the Chattahoochee National Forest.

White County

- The Old Brock Homesite Cache: Gold coins and family silver buried on the Brock homesite, 4 miles northeast of Clermont, close to Route 129 between Gainesville and Cleveland, and near the Chattahoochee National Forest.

- The Trader Cache: $40,000 buried by a trader on his homestead, halfway between Nacoochee and Robertstown.
- The Lost Settler Silver Mine: Located somewhere within 5 miles of Robertstown, near Route 17.

Whitfield County

- The Stage Station Bullion: Gold bullion is hidden in or near the ruins of a stage station near Route 201 and Cahulla Creek, 15 miles south of Chattanooga, Tennessee at the northwest corner of the Chattahoochee National Forest.
- The Cherokee Buried Treasure: One thousand bars of gold in a cave on Rocky Face Mountain, near Varnell on Route 2.

Wilkes County

- The Confederate Treasury: Most of the $10 million in the Confederate Treasury in said to have been buried in different locations in the area surrounding the town of Washington. (Some of the paper currency has been found, but the gold bullion is still missing.)
- The Dionysus Chenault Plantation Hoard: $200,000 in gold was buried near the horse pen of the Dionysus Chenault Plantation, 12 miles northeast of Washington on Route 44.

FLORIDA

There is no dearth of treasure lore along the coast of mainland Florida and in the Florida Keys, where treasure hunting is a major local cottage industry. In the Florida Straits and throughout the Caribbean, the search for sunken Spanish galleons and pirate ships is big business, and a business that often yields big profits. Discovery of such wrecks in the 1980s and 1990s has earned millions of dollars, although the expenses in an underwater search can also climb into seven figures. For the purpose of this book however, we concentrate on treasures located on land, in coastal waterways or in the Everglades.

Spanish buried treasure and pirate treasure account for a major part of Florida's booty, but there are also fortunes lost or hidden here during the Civil War, or during the Seminole Wars of 1816 and 1835–1842. General (and later President of the United States) Andrew Jackson played an important role in the First Seminole War, and his name comes up in the treasure lore as well. In one instance, Jackson hung two traders for selling arms and inciting the Seminoles. Their hoard, which was not recovered (at least at the time), was buried near the junction of two streams near the north edge of Cross City in Dixie County.

On another occasion, Jackson and his troops were pursuing a group of Seminoles who had hidden seven horse loads of gold and silver coins in a swampy area, now called "Money Pond" locally, which is near Neal's Landing, southwest of Fernandina Beach, near the Georgia state line in Nassau County.

Curiously, the name "Billy Bowlegs" occurs twice in nineteenth-century Florida lore, as there were two persons who shared that appellation. The first Billy Bowlegs was the noted Seminole leader and raider of the 1830s, who plundered a great deal of loot from both white settlers and other Seminoles. The second was an English pirate whose real name is said to have been William Rogers, though we're really not sure.

According to legend, Rogers first appeared in Louisiana around 1810, purchased a plantation, married a Choctaw woman, and had six children. In the meantime, he is said to have sailed with Jean Lafitte at the Battle of New Orleans in 1814, and to have become a pirate at about the same time. He was active through the 1830s, having established his base around Santa Rosa Island in the Florida panhandle. In one spectacular incident, he outmaneuvered a pursuing British warship in the Santa Rosa island sandbars, but was forced to scuttle his own ship to keep the plunder away from the Brits. Unfortunately, the gold wound up too deep for him to retrieve. He died in 1888 at the age of ninety-three without ever having recovered this loot, and he left other caches throughout the Florida panhandle area between Pensacola and Panama City, particularly on the offshore islands and of Santa Rosa County.

One of the oldest Florida treasure legends tells us that in about 1550 the Indians killed the survivors of a Spanish shipwreck and buried a large hoard of gold coins and jewelry at Turtle Mound, seven miles south of Coronado Beach near what is now New Smyrna.

Four decades earlier, in 1513, the Spanish explorer Ponce de Leon came here in search of the most treasured of treasures, the Fountain of Youth. He never found the fabled fountain, but his name is still associated with Ponce de Leon Springs in Volusia County, which is not the Fountain of Youth. Several treasures are reported to have been stashed near here, such as the one at the ruins of a Spanish sugar mill which was destroyed by the Seminole. In the 1720s, Spaniards, chased by Indians, buried a chest of silver coins within three hundred paces of Ponce de Leon Springs.

One of Florida's most notorious pirates was José Gaspar, known as "Gasparilla the Pirate." It has been estimated that between 1784 and 1821 he plundered more than $5 million (worth many times that today), of which only about $17,000 has been recovered. In one instance, Gasparilla was being chased by several ships when he went ashore on Anastasia Island near St. Augustine and buried a chest containing $50,000 in gold coins near a large oak tree.

Amelia Island was a notorious pirate's lair and several small caches of gold doubloons were found in the 1930s. Among those still waiting for a finder are Luis Avery's booty near the southern end of the island, Stede Bonnet's hoard of church plate near Fort Clinch, Blackbeard's chests buried in the vicinity of Fort Clinch and two large pirate chests buried in a patch of palmettos about 1.5 miles from the south end of the island, two hundred paces from the beach.

Tampa Bay was home to buccaneers long before it was home to the Buccaneers of the National Football League, and with the former came hoards of plunder. Two longboats filled with pirate booty are supposed to be buried close to the airport along Sweetwater Creek near Rocky Point on the east side of Tampa Bay. There have been several tales of buried treasure on Christmas Island, near the mouth of Tampa Bay, including the hoards stashed by Gasparilla, the $50,000 in paper currency buried in the 1920s by a bank robber and the Prohibition-era rum runners' caches.

There are several rumored treasures near the Courtney Campbell Causeway connecting Tampa and Clearwater across

Tampa Bay. A rich farmer buried a treasure chest in the center of the triangle formed by three oak trees on the top of Pierce's Bluff. Nearby Copper's Point, north of the causeway, was a pirates' lair and the many markings on rocks in the area might be directions to the buried treasures.

Not all the treasures borne to Florida's shores aboard ships were pilfered by pirates. A great deal of it was hidden after being salvaged by sinking or sunken ships. An English ship was wrecked near Mayport in 1784 and the survivors buried four chests of gold coins. When they reached St. Augustine two days later, they were hanged as spies by the Spanish and no one ever found the cache.

Many of the caches in the Florida Keys were buried by shipwreck survivors but some were also the hiding places of the salvagers sent to recover cargo and treasure. They were putting away a little something for their old age. Treasures have been found on Key Largo, Grassy Key, "Treasure Beach," and elsewhere.

There have been literally hundreds of shipwrecks in the Florida Keys through the years, and the island of Indian Key was used as the main salvage area for the Spanish and, after 1810, by American wreckers, all of whom had the habit of burying their hoards. During the nineteenth century, several of the richest Indian Key wreckers are known to have had hoards hidden on the island and these have not been found. In 1840, during the Second Seminole War, Seminole raiders lured the US Army away and attacked, killing nearly everyone and burning the buildings. Some of the salvaged treasure was stolen, some lost, and much was never found.

The Everglades have many buried treasures. The most well documented is the story of the Confederate paymaster pursued by Union troops, who buried a million-dollar payroll ($200,000 in gold and the remainder in paper currency). He wrote, "Chased by the enemy, we buried our payroll at a point in the Everglades at the junction of two creeks, where the land rises like a camel's back. The money is buried in the west hump of the rise."

The site is somewhere between Alligator Alley and Route 41 near the Seminole Reservation in Hendry County.

There are at least three Civil War treasures in or around the Steinhatchee River in Taylor County. There was $500,000 in silver bullion recovered from a wreck at the mouth of the river and buried nearby. Blockade runners buried $140,000 in gold coins on the riverbank near the river's mouth and Union soldiers buried gold coins, taken from the federal capture of Cedar Key and other Confederate strongholds, about five miles up the river.

Some treasures were actually buried by their rightful owners. There is the story of Richard Crowe's "inheritance." When Crowe died in 1894, his will stated that he had buried $60,000 in gold coins somewhere on his land near St. Augustine. The property was dug up, but the treasure was never found. Here it would be useful to consult 1894 St. Augustine newspapers for more details.

Not all pirate plunder was taken at sea. The John Ashley Gang were bank robbers who operated in Florida from 1915 to 1926, using Canal Point on the southern tip of Lake Okeechobee as their base of operations. Supposedly they buried most of their loot, including $110,000 in gold from their last bank robbery, at Canal Point.

Sometimes there is treasure that is extraordinarily hard to find, but when found, it is simply there to be picked up. And then there are those treasures whose location is known with absolute precision, but which are impossible to get. There is nothing more frustrating (well, almost nothing) than a fortune that can be seen but not touched. One such treasure is in a pool formed by a cold water spring near Ponce de Leon Springs in Volusia County. An iron chest lies there, easy to see. The last we heard, divers had been unable to recover it, and several attempts to use drag lines have failed.

Then there is the so-called "Mystery Chest," which lies in a swamp, surrounded by quicksand and appearing only during dry seasons. One attempt at recovery involved a helicopter and grappling hook, but the down draft from the copter's blades made the chest sink out of sight. For those wishing to try another approach, it is located between the Indian River and the Atlantic Ocean, close to State A1A on the outer bank, eight miles south of Vero Beach.

Broward County (The Everglades)

• The Confederate Payroll: A million-dollar payroll ($200,000 in gold and the remainder in paper currency) buried in the

Everglades "at the junction of two creeks, where. the land rises like a camel's back . . . in the west hump of the rise" (possibly between Alligator Alley and Route 14, near the Seminole Reservation and possibly in Hendry or Collier County).

Charlotte County

• The Treasure of Carlos of the Calusas: In the vicinity of Charlotte Harbor.

Clay County

• Spanish Buried Treasure: Green Cove Springs.

Dade County

• Pirate Treasure: In North Miami.
• The Pirate Treasure of Coconut Grove: In Miami.
• The Pirate Treasure on Snake Creek: Located on Snake Creek in Ojus.

Dixie County

• Andrew Jackson's Justice: The money belonging to two traders hung for selling arms and inciting the Seminoles in 1816 is buried near the junction of two streams near the north edge of Cross City, on US 19 and US 98.
• The Bahama Trader's Treasure: In the vicinity of Cross City.

Duval County

• The English Treasure of 1784: Four chests of gold coins buried after a shipwreck that occurred close to Mayport, near Jacksonville.
• The Lost Pirate Treasure: In the vicinity of Mayport.

Escambia County

- The Lost Treasure Chests: Pirate treasure in or near Pensacola.
- The Spanish Cove Treasure: In the vicinity of Pensacola.

Franklin County

- Billy Bowlegs's Hidden Millions: $5 million in gold and silver coins hidden near Franklin.
- Billy Bowlegs's Treasure: At Bald Point south of Ochlockonee Bay.
- Billy Bowlegs's Treasure: Located on Dog Island off Carrabelle.
- The Copeland Pirate Treasure: Three kegs of treasure buried in Money Bayou, south of Route 98 and west of Apalachicola in Franklin or Gulf County.

Hillsborough County (East Tampa Bay Area)

- Pirate's Treasure of Plant City: At Plant City.
- The Sweetwater Creek Cache: Two longboats filled with pirate booty buried near the International Airport, along Sweetwater Creek near Rocky Point on the east side of Tampa Bay.
- The Treasure of Turtle Mound: At Turtle Mound.

Hillsborough and/or Pinellas County

- The Courtney Campbell Causeway Treasures: One treasure chest is buried in the center of the triangle formed by three oak trees on the top of Pierce's Bluff, and at nearby Copper's Point, north of the causeway, many markings on rocks may be directions to the buried treasures. [The Courtney Campbell Causeway (Route 60) connects Tampa and Clearwater, Hillsborough and Pinellas counties, across Tampa Bay.]
- Christmas Island Treasures: Hoards left by the pirate Gasparilla, the $50,000 in paper currency buried in the 1920s

by a bank robber and the Prohibition rum runners' caches are on or adjacent to, Christmas Island, near the mouth of Tampa Bay.

Indian River County

• The Mystery Chest: Surrounded by quicksand, and appearing only during dry seasons, it lies in a swamp on the outer bank between the Indian River and the Atlantic Ocean, 8 miles south of Vero Beach.

Lee County

• Black Caesar's Treasure: Reported in several locations, including Caesar's Creek, Key West, and Sanibel Island.
• Gasparilla's Main Base: Upper Captiva Island off Fort Meyers.

Levy County

• Jean Lafitte's Treasure: According to legend, Lafitte buried treasure near three large oaks at Fowler's Bluff near Chiefland, 15 miles up the Suwannee River.

Marion County

• The Don Felipe Family Treasure: Gold coins and silver buried on the family plantation 2 miles northwest of Ocala, before Don Felipe himself was killed by Indians and his plantation destroyed.
• De Soto's Buried Treasure: In the vicinity of Silver Springs.

Monroe County (The Florida Keys)

• Hidden Shipwreck Salvage:
  Key Largo
  Grassy Key
  Indian Key
  Treasure Beach

- Pirate Treasure:
  Anclote Key
  Boca Chica (The Pirate Demons' Treasure)
  Elliot Key
  Key West (a possible site of Black Caesar's Treasure)
  Old Matecumbe Key
  Panther Key (Gasparilla pirate treasure)
  Pigeon Key
  Siesta Key
- Buried Mexican Treasure:
  Grassy Key

Nassau County

- Luis Avery's Booty: Located near the southern end of Amelia Island.
- Stede Bonnet's Hoard: Church plate hidden near Fort Clinch on Amelia Island.
- Blackbeard's Chests: Two large pirate chests buried near Fort Clinch, in a patch of palmettos about 1.5 miles from the south end of the Amelia Island, 200 paces from the beach.
- Captain Kidd's Treasure: Fernandina Beach.
- The Money Pond: Seven horse loads of gold and silver coins hidden by Seminole people in a swampy area near Neal's Landing, southwest of Fernandina Beach, near the Georgia state line.

Palm Beach County.

- Blackbeard's Buried Treasure: In or near Boca Raton.
- The John Ashley Gang's Loot: Hidden between 1915 and 1926, most of their loot, including $110,000 in gold from their last bank robbery, in near Canal Point, on the southern tip of Lake Okeechobee.

Pinellas County (see also Hillsborough County)

- The Big Island Treasure: Native Americans buried their treaty payment money, seven casks of gold and silver coins at Big Island on the western side of Tampa Bay.

- War of 1812 Treasure: During the war, the British buried a pay chest on a site now under the Vinoy Park Hotel in St. Petersburg.

St. John's County

- Gasparilla's Treasure: A chest containing $50,000 in gold coins hidden near a large oak tree on Anastasia Island near St. Augustine.

- Richard Crowe's Inheritance: $60,000 in gold coins buried somewhere on Crowe's property near St. Augustine before 1894.

St. Lucie County

- The Ashley Gang Loot: $25,000 buried near St. Lucie Inlet.

Santa Rosa County

- Billy Bowlegs's Pirate Treasure: $3 million hidden in a cavern below Fort San Carlos at Pensacola.

- The Pirate Treasures of Billy Bowlegs: Hidden on Santa Rosa Island, part of which lies in Okaloosa County.

Taylor County

- The Civil War Treasures of Steinhatchee River: $500,000 in silver bullion recovered from a wreck at the mouth of the river and buried nearby; $140,000 in gold coins buried on the riverbank near the river's mouth; and gold coins, taken from the Federal capture of Cedar Key and other Confederate strongholds, buried about 5 miles up the river.

Volusia County

- The Turtle Mound Massacre Treasure: A large hoard of gold coins and jewelry from a 1550 shipwreck buried at Turtle Mound, 7 miles south of Coronado Beach near New Smyrna.

- The Spanish Sugar Mill Treasure: Located near the ruins of a Spanish sugar mill which was destroyed by the Seminoles, on the east side of Route 17 northwest of Ponce de Leon Springs.

- The Pirate's Iron Chest: Visible in a pool formed by a cold water spring near Ponce de Leon Springs, Volusia County.

- The Ponce de Leon Springs Spanish Treasure: A chest of silver coins buried in the 1720s within 300 paces of Ponce de Leon Springs, 7 miles north of Deland.

Wakulla County

- Lewis Leland's Pirate Treasure: In the vicinity of St. Marks.

Walton County

- Billy Bowlegs's Cache: Located near the mouth of a small river that empties into Choctawatchee Bay north of Fort Walton Beach.

ALABAMA

In Alabama, the treasure lore includes both Civil War and pirate treasure. In the case of the latter, the Gulf Coast of Baldwin and Mobile counties is particularly rich. This includes the easternmost of the caches attributed to the legendary buccaneer Jean Lafitte, who was particularly active in Louisiana and the Texas Gulf Coast.

Farther north, in Limestone County, the Civil War is recalled in a pair of metal boxes stuffed with $100,000 in gold and silver that were in boxes and dumped into a swamp four miles north of Athens in 1865 by Confederate troops.

Baldwin County

- Henry Nunez's Hoard: $200,000 buried by Henry Nunez near the ruins of his house, on the Alabama side of the Perdido River, near where US 90 now crosses the river, about 16 miles northwest of Pensacola, Florida.
- Pirate's Treasure: An unidentified pirate's treasure is buried near Bay Minette.
- The Pirate Treasure of Old Fort Morgan: At Fort Morgan near Mobile Point.

Limestone County

- Lost Confederate Treasure: Over $100,000 in gold and silver packed in two metal boxes and dumped into a swamp 4 miles north of Athens.

Mobile County

- Jean Lafitte's Buried Treasure: $80,000 in gold coins buried on a beach in Bayou la Batre, south of Mobile.

Pickens County

- The Iron Treasure Chest: In the vicinity of Gordo.

Tallapoosa County

- The Lost Civil War Cache: In the vicinity of Dadeville.

MISSISSIPPI

In addition to the pirate and Confederate caches which are typical in the Gulf states of the former Confederacy, the Magnolia State has several other interesting tales of hidden troves in its treasure lore. In September 1850, the steamboat *Drennan White*, carrying $100,000 in gold coins, sank in the Mississippi River near what was the Ancil Fortune Farm, fifteen miles south of Natchez. The river later changed course and the shipwreck is now on dry land four miles west of Route 61.

Before dying in 1945, an old miser named Hiberly hid caches of gold coins on his farm two miles northeast of Lumberton; gold and jewelry were buried somewhere on the Bond farm near the eastern outskirts of Natchez; the millionaire T. P. Gore buried $400,000 in gold bars and coins near his mansion in Calhoun City; and Native Americans robbed an oxcart carrying a load of silver bars up to Louisville, burying the loot near the foot of a cliff near Williams.

During the early days of the Great Depression as news of the October 1929 stock market crash spread and banks began to fail, many people began burying their money rather than trust banks. Of course, it was the hysterical withdrawals that had caused the banks to fold in the first place. In Mississippi one old recluse buried $18,000 in gold coins in the town park at Doddsville, and Zack Goforth buried $30,000 on his farm near Little Rock, and it was probably never recovered.

The notorious thief Sam Mason hid a large treasure at the Devil's Punchbowl on the Mississippi River south of Natchez and buried $75,000 in gold and silver coins in a cemetery at Little Sand Creek near Rocky Springs on the Big Black River. The Mason Gang also is said to have dropped a large pot filled with gold coins and bullion into an artesian well on the Robert Dove farm south of Hamburg on the northern edge of the Homochito Forest.

The Gulfport area is home to perhaps the largest concentration of buried treasure stories in the state. There are literally hundreds of caches of family valuables hidden in and around the old plantations in the vicinity.

Of the fifty states that we contacted for clarification of laws governing treasure hunting, Mississippi was one of eleven that supplied useful information. The office of their attorney general replied that under the state statutes 39-7, shipwrecks and buried treasure that are designated as state landmarks are the sole property of the state, but that salvage permits can be obtained in certain cases. Landmarks include "sites, objects, buildings, artifacts, implements, and locations of archeological significance." Inquire locally, consult your attorney, or write to

State of Mississippi
Office of the Attorney General
450 High Street
Post Office Box 220
Jackson, Mississippi 39205-0220

Adams County

- Pirate's Treasure: In the Natchez area.
- Sam Mason's Hidden Treasure: A large treasure at the Devil's Punchbowl on the Mississippi River south of Natchez.
- The White Horse Tavern Treasure: In Natchez.
- The Bond Farm: Gold and jewelry were buried somewhere on the Bond Farm, near the eastern outskirts of Natchez.

Calhoun County

- T. P. Gore's Hoard: The millionaire buried $400,000 in gold bars and coins near his mansion in Calhoun City.

Choctaw County

- The Ox Cart Robbery: Native Americans buried their plunder near the foot of a cliff near Williams on Route 15.
- Pirate's Treasure: A pirate supposedly buried $100,000 in gold coins along the Big Black River near Mathison on Route 82.

Clairborne County

- Sam Mason's Cemetery Hoard: $75,000 in gold and silver coins buried in a cemetery at Little Sand Creek near Rocky Springs on the Big Black River.

Franklin County

- The Mason Gang's Lost Gold: A large pot filled with gold coins and bullion in an artesian well on the Robert Dove Farm, south of Hamburg at the north edge of Homochito Forest.

Greene County

- The Gaines Farm Hoard: Gold and silver coins buried on the Gaines farm near Leakesville, on the west side of the Chickasawhey River adjacent to Routes 57 and 63.
- The Spanish Treasure: In the vicinity of McLain.

Hancock County

- The Copeland Gang's Cache: In Catahoula Swamp.
- The Pirate's House: Legend says there is treasure buried in or around such a house in the center of Bay St. Louis.
- The Napoleon Church Gold: $80,000 in gold coins buried near where the Napoleon Church stood, in a grove of oak trees 100 yards from the Pearl River near Waveland.

Harrison County

- Patrick Scott's Hidden Treasure: In or near Biloxi.
- The Lost Gulfport Caches: Family valuables hidden in and around the old plantations in the vicinity of Gulfport.
- Captain Dane's Treasure: $200,000 in an oak grove on an old plantation near Pass Christian, near Route 90.

Jefferson County

- Joseph Hale's Cache: $70,000 in stolen gold and silver coins buried near Fayette, on Route 28 and 33.
- A Mason Gang Cache: $25,000 in gold coins buried near Tillman, north of Natchez.

Jefferson Davis County

- The Bank Robbery Money: $80,000 buried in a field near Prentiss on Route 13.

Jones County

- The Copeland Gang's Hoard: At Big Creek, near Laurel on US 59.

Lamar County

- Hiberly's Hoards: Caches of gold coins on the Hiberly farm, located 2 miles northeast of Lumberton on Route 11.

Lauderdale County

- A Copeland Gang Cache: Loot buried in the ghost town of Arundel Springs at the junction of Sowashee and Okitibbee Creeks south of Meridian on Route 45.

Lawrence County

- Cooper's Creek Bridge: Cache buried on a hill near the Cooper's Creek Bridge and Robinwood Road between Robinwood and Monticello.

Leflore County

- The Mayor of Greenwood's Treasure: His honor hid two large kegs with valuables along the banks of the Yalobusha River near Greenwood.

Lincoln County

- Native American Hoards: Some of the US government payments, made during the mid-nineteenth century in gold coins, were buried a half mile north of Bogue on the banks of the Bogue River.
- Chicken Willy Smith's Tavern Treasure: An unknown sum is buried under Chicken Willy Smith's Tavern near Caseville on Route 550.

Marion County

- The Lost Barrels of Broad Street: Two barrels of gold coins and silverware were buried in a vacant lot on Broad Street in Columbia.

Marshall County

- The Union Paymaster's Treasure: A Union paymaster buried $80,000 in gold coins within sight of the railroad station in Holly Springs.

Newton County

- Goforth's Hoard: Zack Goforth buried $30,000 on his farm, near Little Rock on Route 15.

Panola County

- Civil War Treasure: Gold and silver coins buried in Como.

Pearl River County

- The Lost Treasure of the Catahoula Plantation: Gold coins and jewelry hidden along the banks of Catahoula Creek, near the manor house on the Catahoula Plantation, a half mile east of Picayune.

St. Louis County

- Bandit Treasure: In the vicinity of St. Louis.

Stone County

- The Lost Gold of the Copeland Gang: $30,000 in gold coins hidden in the Catahoula Swamp near Wiggins and Route 49.

Sunflower County

- Doddsville's Lost Civil War Treasure: In Doddsville.
- Doddsville's Town Park Cache: $18,000 in gold coins was buried in 1929 in the town park at Doddsville.

Warren County

- John Murrell's Loot: $400,000 buried near Blakely, just north of Vicksburg.
- The Pickett Plantation Treasure: $200,000 is buried near the mansion on the Pickett Plantation on the north edge of Vicksburg.

Wilkenson County

- The Steamboat *Drennan White*: The vessel, carrying $100,000 in gold coins, wrecked on the Mississippi River near the Ancil Fortune Farm, 15 miles south of Natchez and 4 miles west of Route 61, in a place that is now dry land.

Yalobusha County

- Dr. John Young's Cache: He buried all his wealth in an iron pot in his yard on Young Street in Walter Valley.

TENNESSEE

A battleground in the Civil War as well as in the Indian Wars of the early nineteenth century, the Volunteer State has a rich treasure lore. There is a Lost Indian Silver Cave Mine on Cumberland Mountain near Crossville and Lost Cherokee Gold Mines in the Bald Mountains.

During the Civil War, Union troops buried between $1 and $2 million in plunder along Owl Creek, two miles north of Lexington; fleeing Confederates hid treasure near Cove Creek Cascades northwest of Gatlinburg and a $60,000 Confederate Payroll is buried on the west side of Lake Barkle near Bear Springs and over a million in gold hidden by the outlaw Jon

Murrell near his stone house three miles south of Dancyville. Some Cherokee raiders are said to have buried $500,000 in bullion and coins in an Indian mound near Log Mountain, six miles northeast of Luttrell.

Cocke County

- The Touhy Gang Treasure: In the .vicinity of Newport.

Coffee County

- Cefe Wenton's Cache: A "fortune" in gold coins buried on Wenton's farm, near Hillsboro on Route 41.

Cumberland County

- The Lost Indian Silver Cave Mine: Located on Cumberland Mountain near Crossville.

Dover County

- The Confederate Payroll: $60,000 in gold and silver coins buried on the west side of Lake Barkle near Bear Springs.

Fayette County

- John Murrell's Outlaw Gold: $1 million buried near his stone house, 3 miles south of Dancyville on Route 76.
- The Wolfe Creek Booty: Outlaw Joseph Thompson Hare hid his booty on the banks of Wolfe Creek outside Rossville.

Grainger County

- The Cherokee Raiders' Gold: $500,000 in bullion and coins buried in an Indian mound near Log Mountain, 6 miles northeast of Luttrell.

Greene County

- The Lost Cherokee Gold Mines: Around Big Butte, 10 miles southeast of Greenville in the Bald Mountains, near Route 411.

Henderson County

- The Union Troops' Cache: Between $1 and $2 million in plunder buried along Owl Creek, 2 miles north of Lexington near Route 20.

Houston County

- John Winter's Cache: He buried a cache of gold coins and silver plate on his farm, 2 miles north of Erin near the Cumberland River and Route 13.

Marion County

- Civil War Loot: Buried by Union troops outside Monteagle in a cave on Monteagle Mountain.

Morgan County

- Lost California Gold: A returning California gold miner buried his bullion and coins on the bluff overlooking Big Clear Creek, west of Wartburg near Route 62.

Sevier County

- Cove Creek Cascades Treasure: Fleeing Confederates hid treasure near Cove Creek Cascades northwest of Gatlinburg off Route 441.
- The Lost Delosie Silver Mine: Somewhere in the Great Smoky Mountains near the North Carolina State line and Gatlinburg off Route 441.

Stewart County

- The Fort Henry Treasure: Buried among the ruins of Fort Henry on the east side of Kentucky Lake.

KENTUCKY

The hills and hollows of the Bluegrass State had always provided useful hiding place for the valuables of Native Americans, and through the years, Civil War combatants, paranoid misers and others with a thorough distrust of banks. There is a sack with $3,000 in gold coins waiting to be found in the area of Pilot Rock near Apex, and a large treasure of gold and family silver taken from several white settlements by the Cherokees that was hidden near Winchester.

Over a century ago, the McNitt emigrant party buried all of their valuables before being attacked by Indians. The few survivors of the subsequent battle couldn't find most of the cache and it probably still lies well hidden in the Levi Jackson Wilderness State Park along the Little Laurel River near London.

Roger Barrell buried about $200,000 in gold coins on his farm near Steff, and in the 1930s, a successful bootlegger buried $4 million in gold coins and paper currency along the Ohio River opposite Cincinnati and near Covington. James Langstaff buried $75,000 in gold coins under his hardware store in Paducah, in 1881 Jack Neal buried $200,000 in gold and silver coins in an orchard on his farm east of Hueysville, John Swift hid $150,000 in silver bullion and ore in a cave on Pine Mountain north of Fishtrap Lake, and Micojah Harpoe and his brothers are said to have hidden gold throughout the state. There is even a reported Jesse James stash near Russelville.

Christian County

- The Pilot Rock Cache: A sack with $3,000 in gold coins buried in the area of Pilot Rock near Apex and Route 189.

Clark County

- Cherokee Plunder: Gold and family silver taken from several white settlements hidden near Winchester on Route 627.

Fayette County

- The Alleghan Hall Treasure: At or near Alleghan Hall, Lexington.
- The Coldstream Stud Treasure: In or near Lexington.

Floyd County

- Jack Neal's Buried Treasure: In or near Hueysville.

Grayson County

- Roger Barrell's Hoard: About $200,000 in gold coins buried on what was Barrell's farm near Steff on Route 62.

Greenup County

- Greenup's Lost Indian Silver Mine: Located near Greenup on the Ohio River off Route 23.

Hart County

- The Horse Cave Caches: Gambler Anthony Caccoma buried a number of caches in and around Horse Cave on Route 218.

Henderson County

- A Harpe Brothers' Cache: Gold coins buried in a cave above a stream on Harpe's Head Road 10 miles south of Henderson.

Kenton County

- Prohibition Gold: $4 million in gold coins and paper currency buried along the Ohio River opposite Cincinnati near Covington.

Laurel County

- The Little Laurel River Cache/The McNitt Party Cache: Personal valuables located in the Levi Jackson Wilderness State Park along the Little Laurel River near London off US 75.

Letcher County

- Lost Indian Mines: Several silver and gold mines were abandoned in the area of the Pine Mountain Range near Kings Creek off Route 160.

Logan County

- The Jesse James Loot: $50,000 in gold taken from a Russellville bank in 1868 and buried on the outskirts of town.

McCracken County

- James Langstaff's Hoard: $75,000 in gold coins buried under what was once his hardware store in Paducah.

Pike County

- Jack Neal's Treasure: $200,000 in gold and silver coins buried in 1881 by Jack Neal in an orchard on his farm east of Hueysville, 5 miles west of Pikesville.
- John Swift's Treasure: $150,000 in silver bullion and ore hidden by the miner in a cave on Pine Mountain, on the north side of Fishtrap Lake.

Webster County

- A Harpe Brothers' Treasure: $300,000 in gold coins at Harpe's Head near Tilden on Route 56.
- The Harpe's Head Road Treasure: Reported near Dixon, this may or may not be an alternate telling of the previous entry as Tilden and Dixon are only about 8 miles apart.

Wolfe County

- The Red River Lost Silver Mine: Located on the south side of Daniel Boone National Forest near Red River Gorge.
- The Pottsville Gorge Treasure: $500,000 in silver bars and a lost silver mine are located on either Lower or Upper Devil's Creek near Pottsville Gorge and Campton on Route 15.

Yette County

- William Pettit's Fortune: $80,000 in gold coins buried somewhere on what used to be Pettit's 2,000 acre farm 3 miles south of Lexington.

ARKANSAS

With an early history that included visits by and parleys between Spanish conquistadors, French traders, and Native American nomads, Arkansas was an important crossroads of culture and commerce. With the latter came thievery and the need to hide both stolen property and property that was meant to be kept from being stolen. The Spanish also stayed long enough to work the modest silver mines that still exist, although their location is uncertain.

Benton County

- The Spanish Treasure Cave: Somewhere in Benton County.

Chicot County

- The Outlaw Treasure of Stuart's Island: Located near Lake Chicot.

Crawford County

- The Buried Treasure of White Mountain: Somewhere on White Mountain.

Izard County

- Madre Vena Cave Treasure: In the vicinity of Pineville.

Montgomery County

- The Lost Spanish Silver Mine: Somewhere in Montgomery County.

Ouachita County

- The Lost Spanish Mine: Somewhere in or near Ouachita County.

Pope County

- The Lovely Brothers' Treasure: Located on Norristown Mountain near Russellville, is the cache of brothers named Lovely, rather than brothers with another name who simply happened to be lovely.

Scott County

- The Lost Louisiana Gold Mine: In the vicinity of Waldron.

Searcy County

- Hernando De Soto's Lost Silver Mine: Somewhere in or near Searcy County.
- Hermit Tabor's Lost Mine: Somewhere in or near Searcy County.

Sevier County

- The Lost Mine of the Cosatot: In the vicinity of Gillham.

Washington County

- William Flynn's Buried Treasure: In the vicinity of Fayetteville.

Yell County

- The Lost Caddo Indian Mine: In the vicinity of Dardanelle.

LOUISIANA

Once the gateway to the vast French Empire that spread across a quarter of what is now the continental United States, Louisiana has had a long and colorful history. New Orleans, the state's largest city, is also the port at the mouth of North America's most extensive river system and the heart of the largest French-English bilingual metropolis in the United States. New Orleans and the surrounding Bayou country were physically detached from the rest of the United States until causeways were built in the early twentieth century and developed a culture that owed more to Cajun, Creole, and Afro-Caribbean influences than to that of the Anglo-American mainstream.

The bayou country was also traditionally friendly to the pirates and buccaneers of the Caribbean, and it was here that the great pirate prince Jean Lafitte had his headquarters, and where he is thought to have hid much of the vast fortune that he stole from British and Spanish freighters and gold ships during the first quarter of the nineteenth century. His base was on the island in La Fourche Parish once known as Barataria, which is now known as Grand Terre or Grand Isle. Lafitte sailed against the British under an American during the War of 1812, earning himself a pardon from President James Madison. Lafitte did not, however, give up piracy, which was one of the conditions of his pardon. Instead, he simply shifted his operations center to Galveston Island, Texas.

The Louisiana bayou country is still full of Lafitte stories. He is said to have murdered his victims and left their bones on Pecan Island, to have had a house near the site of the Barbe House on Shell Beach Drive in Lake Charles, and a large cache of gold and silver coins of Lafitte's era were discovered in 1923 at the salt dome known as Jefferson Island. When Lafitte disappeared in 1825, many legends circulated about the possible locations of his caches in Louisiana and Texas. There are even stories that his ghost still haunts the bayous.

However, while Lafitte's caches are the cornerstone of Louisiana treasure lore, there are other important caches, stashes, and troves. There is $90,000 in gold buried near Ivan, $50,000 on the Conrad Plantation near Baton Rouge area, fruit jars filled with ten- and twenty-dollar gold coins buried near Baskins and a chest of gold never recovered after the Civil War that is buried on the old Metoyer Plantation near Natchitoches.

One of the potentially richest treasure sites in Louisiana is Honey Island near Pearl River in Tammany Parish. There are rumors of at least five different caches, including outlaw Dick Calico's $2.5 million buried in the middle of the island and Pierre Rameau's $450,000 in gold bullion.

Another staple of Pelican State treasure lore are the stories of great hoards buried around the sprawling manor houses of antebellum plantations. The Parlange family buried $400,000 in gold and silver during the Civil War on their plantation in Pointe Coupee Parish, there is $400,000 buried on the Bonafice Plantation in St. John the Baptist Parish, and Hippolyte Chretien buried several chests containing $650,000 in gold and silver coins somewhere on the Chretien Point Plantation in St. Landry Parish.

Walter C. Flowers buried $6 million on his estate between Madisonville and Chinchuba and Colonel Norman Frisby hid two wagon loads of treasure in the marshlands south of the mansion on the Frisby Plantation between Newellton and New Light.

There has always been a strong undercurrent of occult and

voodoo influence in New Orleans, and this has naturally seeped into the lore of lost loot. Marie Lavel, a reputed "voodoo queen," is said to have buried $3 million at her last home on St. Ann Street in the Crescent City.

Of the fifty states that we contacted for clarification of laws governing treasure hunting, Louisiana was one of eleven that supplied useful information. Louisiana also has a useful definition of "treasure." The applicable statute is the Louisiana Civil Code Title XXIII, Occupancy and Possession (Article 3420 Treasure). It states that "One who finds a treasure in a thing that belongs to him or to no one acquires ownership of the treasure. If the treasure is found in a thing belonging to another, half of the treasure belongs to the finder and half belongs to the owner of the thing in which it was found."

In Louisiana, a treasure is defined as "a movable hidden in another thing, movable or immovable, for such a long time that its owner cannot be determined . . . A treasure is not a lost thing, an abandoned thing, or a thing that has no owner. It is a thing hidden in another thing by someone who cannot prove his ownership."

According to Article 3423 of the Louisiana Civil Code of 1870, a treasure is a thing hidden or buried in the earth. According to modern civil law, however, a treasure may be hidden in a movable or in an immovable. Article 3420 deviates from the text of Article 3423 of the 1870 Code and follows the modern approach.

Beauregard Parish

• The Lost Wyndham Creek Mines: Somewhere along the creek there is supposed to be a lost Indian gold mine (*see also* Calcasieu Parish).

Bossier Parish

• The Hubbard Treasure: $90,000 in gold buried near Ivan on Route 160.

Caddo Parish

- The Holt Farm Cache: Several hundred thousand dollars believed to have been buried before 1963, on the farm, 5 miles northwest of Vivian on Route 1.

Calcasieu Parish

- A Jean Lafitte Hoard: Located near Starks, about 4 miles east of the Sabine River on Route 12.
- Jean Lafitte's Shell Beach Treasure: Hidden near the Barbe House on Shell Beach Drive at Contraband Bayou near Lake Charles.
- The Lost Wyndham Creek Mines: Somewhere along the creek near De Quincy, there is supposed to be a lost Indian silver mine (*see also* Beauregard Parish).
- A Jean Lafitte's Buried Treasure: In the vicinity of the Mermenteau and Calcasieu Rivers.

Cameron Parish

- The Shell Bank Cache of Jean Lafitte: Reported to be near a shell bank in Lake Misere, up the Mermenteau River, north of Cheniere Ridge.

East Baton Rouge Parish

- The Conrad Plantation: $50,000 is hidden on the Conrad Plantation near Baton Rouge.

Franklin Parish

- The Evans Cache: Several large fruit jars filled with ten- and twenty-dollar gold coins are hidden on the Evans farm, 3 miles east of Baskins on Route 128.

La Fourche Parish

- Lafitte's Grand Terre Treasure: Possibly located on Grand Isle (formerly Barataria), between Barataria Bay and the Gulf of Mexico.

La Fourche and/or Plaquemines Parishes

- Johnny Gambi's Buried Treasure: Hidden somewhere in the area of Barataria Bay, which touches both parishes.

Grant Parish

- The Gold Miner's Buried Treasure: In the vicinity of Selma.
- The Robber's Treasure of Hull Lake: Located near or in Hull Lake.

Jefferson Parish

- Lafitte's Million Dollar Hoard: Legend holds that Lafitte buried $1 million near the cemetery where he now lies, 2 miles south of Lafitte Village, 21 miles south of Marrero (suburban New Orleans) on Route 45.

Livingston Parish

- Jean Lafitte's Treasures: Buried across the Amite River from the ruins of Galvez Town, 2½ miles south of Oak Grove and 2½ miles west of Port Vincent.

Natchitoches Parish

- The Caches of Grand Encore: During the Civil War, as Union troops approached, many residents buried their valuables along the Red River near Grand Encore.
- The Gambler's Lost Hoard: In 1852 a gambler hid his winnings in a bluff overlooking the Red River near Grand Encore.
- The Murrell Gang's Loot: Up to $20,000 buried in the vi-

cinity of Grand Encore, 4 miles north of Natchitoches on the bank of the Red River.

- The Natchitoches Treasure: A chest of gold, never recovered after the Civil War, had been buried on the old Metoyer Plantation at Natchitoches.
- The Simmons Treasure: A cache of gold hidden on the Simmons House property in Natchitoches.

Orleans Parish

- Queen Marie Lavel's Treasure: $3 million buried by the reputed "voodoo queen" at her last home, on St. Ann Street in New Orleans.
- The Old New Orleans Mint: There are supposed to be several treasures buried on the grounds of the old New Orleans Mint on Esplanade Avenue.
- The Destrehan Plantation Mansion: $470,000 in or around the Destrehan Plantation mansion near New Orleans.

Plaquemine Parish

- Ballowe's Hoard: The pirate Ballowe buried $260,000 in gold and silver near Point a la Hache, on the east side of the Mississippi River on Route 39.

Pointe Coupee Parish

- The Parlange Plantation: The family buried $400,000 in gold and silver during the Civil War on the Parlange Plantation near New Roads.

Quachita Parish

- Confederate Treasure: $20,000 in gold and silver coins at Old Camp Place 10 miles west of Monroe.
- The Old Camp Place Treasure: $150,000 hidden also near Old Camp Place (this may or may not be a variation on the story in the entry above, which is also at Old Camp Place).
- The Limerick Plantation Treasure: At Limerick Plantation.

Red River Parish

- The Red River Treasure: $2 million in gold bullion buried on the banks of the Red River near Coushatta on Route 71.

St. James Parish

- The Indian Mounds Cache: $200,000 buried among Indian mounds near Lutcher on the Mississippi River.
- The Lake Plantation Treasure: In or near the town of St. Joseph.

St. John the Baptist Parish

- The Bonaface Plantation: $400,000 hidden on the grounds of the plantation, on the north bank of the Mississippi River, near Edgard.
- The Buried Treasure of St. Rose: In or near St. Rose.
- The D'Estrahan Plantation Treasure: Located on the grounds of the d'Estrahan Plantation.

St. Landrey Parish

- The Grand Coteau Buried Treasure: In or near Grand Coteau.
- The Chretien Point Plantation: Several chests containing $650,000 in gold and silver coins buried somewhere on the Chretien Point Plantation near Opelousas on Route 190.
- The Fusilier de la Calire Mansion Cache: $500,000 said to be buried in the gardens of this mansion in Grand Coteau on Route 167.

St. Martin Parish

- A Jean Lafitte Treasures: Located on Jefferson Island, 9 miles west of New Iberia on Route 90.
- The Thibodeaux Plantation Treasure: Located on the grounds of the plantation, near Breaux Bridge on Route 31.

- The Pine Alley Plantation: $150,000 is buried on the Pine Alley Plantation near St. Martinsville on Route 31.

St. Tammany Parish

- The Honey Island Treasures: Five different caches on this island near Pearl River, including outlaw Dick Calico's $2.5 million buried in the middle of the island, and Pierre Rameau's $450,000 in gold bullion.
- The Walter C. Flowers Fortune: $6 million buried on the Flowers Estate, midway between Madisonville and Chinchuba.

Tensas Parish

- The Treasure of the Frisby Plantation Ruins: Two wagon loads of treasure worth a million dollars was buried in the marshlands south of the mansion, now in ruins, on the Frisby Plantation north of Route 4, midway between Newellton and Newlight.

Terrebonne Parish

- The Buried Treasure of Caillou Island: Hidden on Caillou Island in Terrebonne Bay.

Vermillion Parish

- Jean Lafitte's Treasures: Plunder is rumored to have been buried on Pecan Island.

Winn Parish

- The Lost Spanish Treasure: Eight mule loads of Spanish gold bullion are buried in a pine forest northeast of Winnfield on Route 156, possibly in the Kisatchie National Forest.

8

THE TREASURES OF THE
GREAT LAKES STATES

Ohio

The treasure lore of the Buckeye State spans the horizon of American history from the mid-eighteenth century when Ohio was a wilderness being fought over by the British and French, to the early twentieth century, when legendary hoodlums warred with the G-men in Prohibition-era shootouts.

During the French and Indian War, a French officer buried plunder on Isle St. George in Lake Erie, fifteen miles north of Sandusky, fleeing British soldiers buried treasures in or near the old Fort Wapatomica on the Muskingum River two miles north of Zanesville and the survivors of a wrecked British warship buried $500,000 at Locust Point, fifteen miles east of Toledo. Riverboat Pirates were later based on a bluff overlooking the Ohio River, one mile northeast of Crown City.

In 1862, a bank robber buried $100,000 in gold bars on the west bank of the Grand River, "three feet deep and thirty paces northwest of a large oak tree on the river bank," two miles from Lake Erie near Fairport Harbor.

One of Ohio's more colorful legends involves the wreck of the Pacific Express. On December 29, 1876, the train plunged from a collapsed bridge near Ashtabula into a river gorge, carrying $2 million in gold bullion into the muddy bottom.

Though his most legendary stash is near the Little Bohemia Resort in Wisconsin, the notorious John Dillinger is said to have buried $1.5 million in loot in a cow pasture on the Pierpont farm near Leipsic, Ohio prior to the famous 1934 crime spree that left him gunned down by FBI agents in Chicago.

Ashtabula County

- The Pacific Express Train Wreck: $2 million in gold bullion in the muddy bottom of a river near Ashtabula and US 20 (check local newspapers for December 29, 1876 and succeeding dates for more details).

Carroll County

- The Sandy River Treasure: $25,000 in gold coins buried on the north side of the Sandy River 1 mile south of Minerva near Route 30.

Columbiana County

- The Morgantown Gang Plunder: Several caches of loot buried by the gang in the 1880s on the west bank of the Ohio River, 2 miles east of East Liverpool near Route 30.

Crawford County

- The Ashland Farm Cache: $25,000 in gold coins buried during the Revolutionary War on the John Ashland farm along the south bank of the Sandusky River, outside Wyandot on Route 231.

Erie County

- The Isle St. George Cache: Plunder buried during the French and Indian War on Isle St. George in Lake Erie, 15 miles north of Sandusky.

Calion County

- The Riverboat Pirates' Loot: Possibly near a bluff overlooking the Ohio River, 1 mile northeast of Crown City near Route 553.

Greene County

- The Shawnee Plunder: Hidden in 1780 along a bend of the Little Miami River near Old Town, 3 miles north of Xenia off Route 68.

Hancock County

- The War of 1812 Lost Payroll: Hidden in or near old Fort Finley, near Finley on US 75.

Lake County

- The 1862 Bank Robbery: $100,000 in gold bars buried on the west bank of the Grand River, "three feet deep and 30 paces northwest of a large oak tree on the river bank," 2 miles from Lake Erie near Fairport Harbor.

Logan County

- Buried British Treasure: In the vicinity of Zanesfield.

Mercer County

- The Lost Army Payroll: Buried in 1790 on the river bank north of Fort Recovery near Route 119.

Morgan County

- The 1924 Bank Hold-up Money: $125,000 buried on the Lisman Farm 2 miles east of Joy near Route 555.

Muskingum County

- The Fort Wapatomica Treasure: Treasures buried by the British in or near old Fort Wapatomica on the Muskingum River, 2 miles north of Zanesville near Interstate 70.

Ottawa County

- The Locust Point Shipwreck: $500,000 in eighteenth century British gold buried at Locust Point, 15 miles east of Toledo.

Preble County

- The Bridge Buried Treasure: Reported near Eaton.

Putnam County

- John Dillinger's Cache: $1.5 million in loot buried in a cow pasture on the Pierpont farm near Leipsic on Route 65.

MICHIGAN

A major industrial center for most of the twentieth century, Michigan has seen great fortunes made and lost. Some of the latter are irretrievable, but others lie waiting for their lucky finder. There is $11 million hidden in or near the House of David Mansion in Benton Harbor. Henry Dansman hid diamonds, gold, and silver coins on his farm between Lake Augusta and Posen; Hodson Burton buried his life's savings of gold coins and paper money on his farm two miles north of Route 12 near Buchanan and Ransom Dopp hid a fortune near his home five miles east of Dowagiac.

As great historical events have crossed the landscape of the Wolverine State, they have often left treasure troves in their wake. In 1812 General Monk, commander of the British garrison, buried a large iron chest filled with army funds near the fort on the southwest tip of Drummond Island in Lake Huron, which is the hiding place of several other reported treasures.

In 1871, after the Great Chicago Fire, looters reportedly buried their booty at Cat Head Point near Northport.

The sites of some possible treasures are within easy reach, but getting to them is very difficult and potentially dangerous. There are several gold mines near Ishpeming on the Upper Peninsula, abandoned and in ruins, that would be easy to find, but entering them is not recommended. Furthermore, they were already abandoned as being unprofitable. It might be fun to go take a look—at a safe distance. Harder to locate would be the fabled Lost Indian Gold Mine in Porcupine Mountain State Park on the upper peninsula near Silver City.

Berrien County

- The House of David Mansion: $11 million is hidden in or near this mansion in Benton Harbor on Route 33.
- Hodson Burton's Hoard: He buried his life's savings of gold coins and paper money on his farm, 2 miles north of Route 12 near Buchanan.

Cass County

- Ransom Dopp's Fortune: He hid his money in or near his home, 5 miles east of Dowagiac on Route 62.

Chippewa County

- The Espanore Island Cache: Spanish gold and silver coins buried on Espanore Island off Drummond Island in Lake Huron.
- The Drummond Island Treasure: $50,000 in gold coins hidden at Potagannissing Bay on the northwestern end of Drummond Island in Lake Huron.
- General Monk's Cache: A large iron chest filled with British Army funds buried in 1812 near the fort on the southwest tip of Drummond Island in Lake Huron.

Leelanau County

- The Great Chicago Fire Loot: Reportedly buried at Cat Head Point near Northport on Route 201 in 1871.

Lenawee County

- Godfrey Watson's Cache: He hid his savings somewhere on his farm, 2 miles north of Tecumseh near Route 50.

Marquette County

- The Lost Jack Driscoll Silver Mine: In the Huron Mountains near Ishpeming.
- The Lost Douglas Houghton Gold Mine: Located near Log Lake and Champion on Route 41.
- The Ropes Gold Mine: Within 3 miles of Ishpeming.
- The Michigan Gold Mine: Within 3 miles of Ishpeming.

Ontonagan County

- The Lost Indian Gold Mine: In Porcupine Mountain State Park near Silver City on Route 64.

Presque Isle County

- Henry Dansman's Cache: He hid his diamonds, gold, and silver coins on his farm, between Lake Augusta and Posen near Route 65.
- Francis Fontenoy's Treasure: In the vicinity of Presque Isle.

INDIANA

Much of the treasure lore in Indiana's history swirls around the great robberies of the late nineteenth century and the Prohibition-era plunder of gangsters and bootleggers. In 1868 the Reno brothers robbed a train, escaping with a safe containing $80,000. With the posse in hot pursuit, they fled to Canada, but they buried the safe south of the border between Rockford (which was their base of operations) and Seymour. In the famous Marshfield Train Robbery, four bandits robbed a passenger train of over $80,000 in gold bullion, coins, and paper currency. The empty safe was found near Marshfield and the thieves were caught and hanged, but the contents of the safe are not known to have been uncovered.

An interesting Prohibition-era treasure trove that does not contain gold or silver is the stash of whiskey worth more than $300,000 that Al Capone hid in a sealed off cave along the shore of Lake Michigan near Michigan City. If whiskey aged over twenty years is good, this might be extraordinary. Capone is said to have had good taste in booze. Meanwhile, John Dillinger is reported to have buried $600,000 on his father's farm near Moorseville.

Adams County

- Prohibition Money: Buried along the banks of the Wabash River where Route 27 crosses the river between Berne and Geneva.

Huntington County

- The Silver Creek Treasure: In the vicinity of Silver Creek.

Jackson County

- The Reno Brothers Loot: A safe containing $80,000 buried in 1868 between Rockford and Seymour.

Knox County

- General John Morgan's Gold: The Confederate general buried $5,000 in gold coins in what is now the George Rogers Clark Historical Park south of Vincennes.

Lake County

- Jim Colosimo's Diamonds: Buried on the outskirts of Crown Point near Route 8.

Laport County

- The Al Capone Gang's Hidden Whiskey: In a sealed off cave along the shores of Lake Michigan near Michigan City.

Marshall County

- Jim Genna's Steel Box of Loot: Buried under a rock pile in a pasture near the intersection of Route 6 and 331 near Bremen.

Martin County

- Absalom Fields' Buried Treasure: In or near Shoals.
- Native American Gold: Gold bullion and figurines buried around 1810 in a cave on the Rocky McBride Bluff, a little north of Shoals near the White River.

Morgan County

- John Dillinger's Loot: $600,000 buried on his father's farm near Moorseville on Route 67.

Perry County

- Lafayette's Lost Treasure: In or near Cannelton.

Spencer County

- The River Boat Treasure: In or near Rockport.

Vigo County

- The Terre Haute Bank Loot: $95,000 embezzled by a bank employee in the 1920s and hidden on a farm near Route 42 on the outskirts of Terra Haute.

Warren County

- The Marshfield Train Robbery Loot: Over $80,000 in gold bullion, coins, and paper currency removed from a safe taken in the robbery and buried in the Marshfield vicinity, possibly near the Illinois state line, 4 miles away.

ILLINOIS

A large chapter in the treasure lore of Illinois involves the plunder stashed by the gangsters and bootleggers that were active in and around Chicago in the 1920s and early 1930s. Much of it remains hidden, with its whereabouts the subject of intense speculation.

In the 1980s, television talk-show host Geraldo Rivera made a big production of opening a Chicago vault on live television that had supposedly been sealed by Al Capone. The fact that it was largely empty embarrassed the vociferous Rivera, but did little to quell speculation about other caches around town. There are rumored to be many hidden treasures in Chicago, buried by well-known figures like Capone and John M. Hoffman, as well as by other infamous gangsters of the 1920s and 1930s. Most of these are thought to be close to the homes occupied by these people at the time. A study of Chicago newspaper obituaries for these individuals would reveal a wealth of details and further leads.

There are other treasures downstate from the Windy City of course. Before his death back in 1833, John Hill buried gold coins in and around his stockade which stood on a site six blocks south of the courthouse in Carlyle. A hundred years earlier, during the French and Indian War, a French paymaster buried $100,000 in gold coins on the Illinois River between La Salle and Ottawa near a place called Starved Rock.

Clinton County

- John Hill's Fort: Gold coins buried in and around the site of his stockade, 6 blocks south of the courthouse in Carlyle.

Cook County

- Chicago's Hidden Hoards: A study of Chicago newspapers, particularly from the late 1920s and 1930s, would provide more information leading possibly to treasures buried by both the famous and the infamous that are rumored to be located throughout Chicago.

Hardin County

- The Cave in the Rock Treasure: $200,000 in gold and silver coins hidden in a cave near the west bank of the Ohio River.

Jo Daviess County

- The Lost Silver Mines of Galena: Native Americans once mined silver in the Galena area.

Madison County

- Vito Giannola's Lunch Buckets of Loot: Mafioso Giannola is said to have hidden twelve lunch buckets filled with treasure on his farm at Horseshoe Lake near Granite City.

Marion County

- Larson's Loot: Gold and silver coins buried by a train robber near Centralia.

Ottawa County

- The Starved Rock Gold: $100,000 in gold coins buried at Starved Rock on the Illinois River between La Salle and Ottawa.

Randolph County

- The Hidden Stagecoach Gold: $44,000 in gold coins taken in a stagecoach robbery in 1809 and buried somewhere near Chester.

Whiteside County

- The Abbot Cache: A large cache of gold coins were buried on the Abbot Farm, off Route 136, near Fulton, one half mile from the Mississippi River, Whiteside County.

Will County

- Sam Anatuna's Cache: $400,000 hidden near Route 66 south of Braidwood.

WISCONSIN

Perhaps the most famous treasure hidden in Wisconsin is the $70,000 in currency hurriedly hidden by gangster John Dillinger in the woods behind the Little Bohemia Lodge, eight miles southeast of Mercer. In April 1934, Dillinger was in the midst of a six-state crime spree, when he and his gang secretly checked into the rustic lodge for a "vacation."

FBI Special Agent Melvin Purvis, acting on a tip, surrounded the place with dozens of tommy-gun–toting agents on April 22. In a shootout that riddled the lodge with machine gun slugs, an FBI man and an innocent bystander were killed, and two of each were wounded. Dillinger and his gangsters escaped unscathed.

Before he left the area, however, Dillinger reportedly buried his loot in several suitcases in the woods behind the Little Bohemia. Because Dillinger had so little time to carry out this task, it is thought that the hiding place is quite close to the Lodge, but many attempts to find it—including a couple filmed for television—have failed to locate it.

A study of the treasure lore of the Badger State unveils many other interesting stories. For instance, in 1832 before being killed in an Indian attack, soldiers buried four saddle-bags of gold coins on the "highest bluff on the Mississippi River," near Fort Crawford and Prairie du Chien.

Native American lore is also a common theme. There is a legend that tells of seventeenth-century French explorers finding a huge cache of golden artifacts in an Indian mound. This stunning trove, like something out of an Indiana Jones movie, was apparently reburied about four miles east of the present Route 12, near the junction of the Black and East Fork of the Black River near Hatfield Village.

In a similar story dating back to 1891, a man named Paul Seifert and his friend found a cave filled with gold and silver Indian artifacts. They set out with a dozen pack horses to recover the relics and were never seen again. The cave is supposed to be around Bogus Bluff, near Gotham and the Wisconsin River.

Bayfield County
(Apostle Islands National Lakeshore, Lake Superior)

- British Buried Treasure: Located on the Apostle Islands in Lake Superior, most of which now constitute the Apostle Islands National Lakeshore.
- Frederick Prentice's Fortune: His fortune is buried near his home, Cedar Bark Lodge, on Hermit Island.
- The British Soldier's Treasure: A chest of silver coins buried on Sand Island.
- The British Treasures of Stockton Island: Several caches on Stockton Island.
- The Hermit Treasure of Wilson Island: Located on Wilson Island.
- The Lost British Hoard: $128,000 in gold coins on York Island.
- William Wilson's Cache: $90,000 buried near his cabin on Otter Island before his death in the 1860s.

Crawford County

- A Lost Payroll: Four saddlebags of gold coins buried in 1832 on the "highest bluff on the Mississippi River," near Fort Crawford and Prairie du Chien.

Dunn County

- The Maxwell Gang's Loot: $40,000 buried near Elk Mound.

Iowa County

- The Coon Rocks Bluff Gold: Several chests of gold stolen from a riverboat, buried around Coon Rocks Bluff near Dodgeville.
- The Indian Treasure Cave: A cave in the vicinity of Arena.

Iron County

- John Dillinger's Cache: $70,000 in currency in several suitcases buried in the woods behind or near the Little Bohemia Lodge, 8 miles southeast of Mercer off US 51.

Jackson County

- Lost Indian Artifacts: A large cache of golden Native American artifacts found in an Indian mound and reburied 4 miles east of Route 12, near the junction of the Black and East Fork of the Black River, near Hatfield Village.

Milwaukee County

- The Expressway Cache: $7,300 hidden where the Thomas Burke house once stood, under or near an expressway, West Allis, Milwaukee.

Oneida County

- R. C. Bennett's Treasure: A reported $1 million in currency buried in or near his home in Eagle River on Route 45.
- John Dillinger's Other Buried Treasure: Located near Rhinelander.

Pepin County

- The Lake Pepin Cache: In 1928 an Illinois businessman hid several caches between Pepin and Stockholm, on the east shore of Lake Pepin.

Polk County

- The Montana Miners' Gold: A wagon load of gold bullion, worth about $200,000, was lost in a quicksand bog south of Balsam Lake, 7 miles northeast of St. Croix Falls near Route 8.

Richland County

- Paul Seifert's Lost Indian Treasure: Gold and silver Native American artifacts found in a cave around Bogus Bluff, near Orion and/or Gotham on the Wisconsin River and Route 133.

Vernon County

- A Jesse James Gang Cache: A safe and two iron boxes of loot buried near Wildcat Mountain, west of Ontario on Route 33.

Walworth County

- Sam Amatuna's Cache: The gangster buried $50,000 wrapped in canvas north of Pell Lake near Route 120.

Winnebago County

- The Buried Swamp Treasure: Somewhere in Winnebago County.
- The Lake Winnebago Treasure: In 1931 a wealthy manufacturer buried $500,000 along the banks of Lake Winnebago near Oshkosh and Route 45.

MINNESOTA

There are a number of interesting stories in Minnesota's treasure lore, with one of the most intriguing being a strange nineteenth-century tale of some children that lived in the small town of Robbin. The children were playing along the banks of the Red River one day, when they found "shiny rocks" the size of apples. They decided to take a few home to show their parents. The adults quickly recognized the "shiny rocks" to be huge gold nuggets. They tried to retrace the children's footsteps, but never found the "thousands of shiny rocks." They are still out there.

Ma Barker was one of the most ruthless gangsters of the 1920s and 1930s. She led a gang of particularly effective bank robbers that included her four sons and her "adopted son," the cruel Alvin "Creepy" Karpis. In 1933, the Barker-Karpis gang turned to kidnapping with the abduction of brewery heir William Hamm. In January 1934, they struck again, this time capturing St. Paul businessman Edward Bremer. The gang netted $300,000 in the kidnappings, but eventually they were tracked down. Ma Barker was killed in a 1935 shootout and Creepy Karpis was nabbed in 1936. He went to jail for thirty-three years, leaving a reported $100,000 in ransom money hidden in a metal box under a fence post somewhere along Route 52 between Chatfield and Rochester. It is not known whether he tried to reclaim it upon his release from federal custody in 1969.

There is a story that a man named Charles Ney buried a fortune in a vault under a brewery on Route 19 in Henderson, Minnesota in 1924. As the story goes, the brewery burned to the ground and Ney couldn't find the vault. We realized that with Prohibition in place, there would be no breweries operating in 1924, so we researched the story further and discovered that the only brewery to operate in Henderson in the twentieth century was that of Hans "John" Enes. It closed in 1920 and never reopened after Prohibition. Maybe Hans Enes lost interest, or maybe the building burned. The ruins may still be visible, and if Ney was, in fact, unable to find the vault as he claims, maybe the treasure is still there.

Faribault County

- Joseph Winther's Cache: He buried his gold coins on his farm, a half mile west of Winnebago on the Blue Earth River near Route 169.

Hennepin County

- The Old Soldier's Home Gold: A settler buried $5,000 in gold coins near the Old Soldier's Home, 2 miles south of Minneapolis on the west side of the Mississippi River.

Kittson County

- The Children of Robbin's Lost Cache: The "thousands of shiny rocks" on the banks of Red River near Robbin on Route 11.

Olmstead County

- Depression Savings: Many prosperous people buried their money near their homes, rather than risk the banks, and legend holds that, for some reason, this was especially prevalent in the Rochester area.
- The Barker-Karpis Gang Ransom Money: $100,000 in a metal box hidden under a fence post somewhere along Route 52 between Chatfield and Rochester.

Pipestone County

- The Jesse James Gang Loot: They are said to have buried $55,000 in gold bullion and coins on Route 30 in or near what is now Pipestone National Monument.

Sibley County

- The Farmer's Cache: $10,000 in gold coins hidden in a grove of trees on a farm 1 mile north of Henderson on the Minnesota River.
- Charles Ney's Brewery Cache: A fortune is said to be in a vault under the ruins of the Enes Brewery on Route 19 in Henderson.
- The Curran Brothers Bank Loot: $40,000 buried on the north end of Mud Lake, near Green Isle on Route 5.

Wabasha County

- The Bootlegger's Lost Cache: He apparently buried a fortune in cash on the banks of the Mississippi River near Lake City just before he was killed in an auto accident on Route 36.

Wadena County

- The Bandits' Gold: A large amount of gold and silver coins buried in the woods west of Wadena on Route 10.

9

THE TREASURES OF THE PLAINS

Missouri

The spectrum of buried treasure in Missouri runs from the sealed and "lost" Native American gold and silver mines of the Ozarks to hiding places used by Confederate raider William Clarke Quantrill during the Civil War, or by bank robbers like Cole Younger and Jesse James a generation later. Indeed, Quantrill's guerrillas stashed an estimated $10 million near a cliff overlooking the Meremac River in what is now Mark Twain National Forest.

In the middle nineteenth century, Missouri was the "jumping-off" point for the great waves of westward migration. Most who passed through going west never came back. Some did. One '49er returning from California buried $60,000 in gold before being bushwhacked and killed at a wagon ford on the Gasconade River at the western edge of Mark Twain National Forest, seven miles west of Lynchburg.

There are two treasures associated with Hannibal, the home town of Mark Twain (Samuel Clemens). Though the great writer himself had nothing to do with them, the favorite son's name seems to be linked with Hannibal in any context. There is supposed to be a ton of gold bars hidden between "Mark Twain's Cave" and the banks of the Mississippi River, and more gold is said to have been hidden in or near McDowell's Cave near Hannibal on Route 61.

The name of Dr. Lynn Talbot is also prominent in Missouri treasure lore. The doctor was murdered when he refused to reveal where he had buried a keg full of gold coins. His "House of The Seven Gables" is north of Barnard on Route 71.

Another interesting tale tells of the paddle-steamer *Francis X. Aubrey* that wrecked on the Missouri River carrying about five hundred barrels of whisky. The river has since changed its course and the wreck is now ashore, buried in a swamp a mile from the river near Parkville and Park College north of Kansas City. It is possible that the barrels are still intact if they have been sealed in the airtight mud.

Barry County

- Native American Gold and Silver Mines: Gold and silver from local mines were sealed in a cave overlooking the White River near Cassville on Route 37.

- Native American Gold and Silver Mines: Gold and silver from local mines were sealed in a cave on the White River near Eagle Rock on Route 86.

Benton County

- The Lake of the Ozarks Treasure: Native Americans buried gold, silver, and jewels in a cave overlooking the Lake of the Ozarks near Warsaw on Route 65.

Cedar County

- The Lost Gold Mine Located on Carpenter Creek: Located on Carpenter Creek near Jerico Springs on Route 97.

Christian County

- The Lost Spanish Treasure: Somewhere in Christian County.

Clay County

- The *Francis X. Aubrey* Whiskey: The paddle-wheel steamer wrecked on the Missouri River carrying about 500 barrels of whisky, but the river has since changed course and the wreck is now buried in a swamp near Parkville and Park College on the north side of Kansas City.

Dallas County

- The Lost Missouri Silver Mine: In the vicinity of a creek near Louisburg on Route 65.

Douglas County

- The Lost Spanish Silver Mine: Somewhere in the Douglas County area.

Greene County

- Confederate Treasure: Hidden on or near Noble Hill, 12 miles north of Springfield on Route 13.
- The Lost Springfield Mines: Gold and silver mines in the vicinity of Springfield on US 44.

Hickory County

- The Lost Brooksie Silver Mine: Somewhere near the southeastern tip of the Lake of the Ozarks near Cross Timbers.

Howard County

- The Kaffer Treasure: A large cache of gold coins is buried near Armstrong on Route 3.

Howell County

- Charles Boucher's Cache: Five hundred silver dollars buried on his farm, in or near his cabin, a mile south of West Plains on Route 63.

- The Hocomo Sinkhole Treasure: $35,000 in gold and silver coins were hidden in a sinkhole on a farm in Hocomo.
- Colonel Porter's Loot: $30,000 in gold coins and silverware hidden almost on the Missouri-Arkansas state line near Lanton on Route 17.

Jaspar County

- The Murdered Cattleman's Treasure: He is thought to have buried $100,000 somewhere on his ranch, 3 miles northwest of Joplin.
- Cole Younger's Cache: He buried a large chest of gold and silver coins near Alba, 7 miles west of Route 71.

Laclede County

- Quantrill's Spoils: The Confederate Raider buried $10 million near a cliff overlooking the Meremac River by Belle Starr's Needle in Mark Twain National Forest.
- Lost California Gold: $60,000 buried west of a former wagon ford on the Gasconade River at the western edge of Mark Twain National Forest, 7 miles northwest of Lynchburg on Route 32.

McDonald County

- The Lost Spanish Gold Mine: Located on the Big Sugar River near Pineville on Route 71.
- The Spanish Miners' Gold: Ore and dust hidden near the creek in Bear Tree Hollow near Lanagan on Route 71.
- The Madre Vena Cave: Spanish gold bullion from adjacent mines is reportedly hidden in the Madre Vena Cave, somewhere in the area of the Ozark Wonder Cave between Noe and Jane, and near the Elk River.
- The Ford Meadow Treasure: $75,000 in gold and silver coins hidden in Ford Meadow near Bethpage on Route 76.

Marion County

- The McDowell's Cave Treasure: Gold was hidden in or near the cave, near Hannibal on Route 61.
- The Mark Twain's Cave Treasure: A ton of gold bars hidden between the cave and the banks of the Mississippi River, near Hannibal.

Nadaway County

- Dr. Lynn Talbot's Cache: A buried keg of gold coins, possibly in or near Talbot's "House of The Seven Gables" north of Barnard on Route 71.

Ozark County

- A Jesse James Cache: $100,000 hidden in a cave near Gainesville on Route 160.

Pulaski County

- The Lost Treasure of Possum Ledge: In the vicinity of Possum Ledge.

Stone County

- The Lost Yokum Silver Mine: Located between the tributaries of the King and St. James Rivers, near Table Rock Lake.
- Lost Spanish Silver Mines: Within a 5-mile radius of Galena.
- The Lost Confederate Silver: $1 million in silver bars hidden along the banks of the James River near Galena on Routes 173 and 176.

Wayne County

- The Lost Treasure of Jesse James: $100,000 in gold bullion and coins buried in the Des Arc Mountains, 8 miles east of Gad Hills on Route 49.

Wright County

- The Spanish Cave Treasure: $600,000 in gold coins hidden in a cave near the Norfolk River, near Hartville on Route 39.

IOWA

The treasure lore of the Hawkeye State includes the plunder left behind by bandits from the James Gang to Bonnie and Clyde. Beneath the wide open spaces of Iowa lie many treasures. Buried on the banks of the Mississippi in Clinton, there is an iron pot of gold coins and stock certificates, and $50,000 in gold coins was buried by train robbers in a five-acre swamp near a creek that enters the Mississippi River at Buffalo, southwest of Davenport on Route 22.

Tom Kelly, a lead miner, concealed $100,000 or more in gold and silver coins on Kelly's Bluff above Second and Bludd Streets in downtown Dubuque, and Mississippi River pirates are said to have hidden plunder in a cave near the river on the south side of Bellevue; at Stone Park, overlooking the Missouri River at Sioux City and at Burris City, on the north side of the Iowa River where it enters the Mississippi.

In the several years leading up to 1832, the great Sauk warrior chief and orator, Black Hawk, along with his warriors buried large amounts of gold at several locations in Iowa. These included the northern side of the Des Moines River in Davis County and in the hills near Demark on Route 103 in Lee County.

Cedar County

- The Ives Brothers' Hoards: Several hoards of money and stock certificates were buried on their farm near Sunbury north of US 80.

Clayton County

- The Army's Lost Payroll: Buried on the banks of Miner's Creek, near Guttenburg on the Mississippi River on Route 52.

Clinton County

- The River Front Park Treasure: An iron pot of gold coins and stock certificates was buried on the Mississippi River in River Front Park at Clinton.

Cott County

- The Train Robbers' Gold: $50,000 in gold coins buried by train robbers in a 5-acre swamp near a creek that enters the Mississippi River at Buffalo, southwest of Davenport on Route 22.

Dallas County

- The Burrow Gang's Loot: $65,000 in gold coins buried near Redfield on Route 6.
- Bonnie and Clyde's Loot: Buried at a camp they used in a wooded area overlooking the Raccoon River, 3 miles north of Dexter on Route 6.

Davis County

- Black Hawk's Gold: Large amounts of gold at two possible locations including the northern side of the Des Moines River, and in Section II, Township 70, Range 12.

Dubuque County

- Tom Kelly's Hidden Fortune: $100,000 or more in gold and silver coins hidden on Kelly's Bluff, above Second and Bludd Streets in downtown Dubuque.

Hamilton County

- The Banditti of the Plains Treasure: Headquartered near the mouth of the Boone River, 2 miles north of Stratford on Route 175, they buried their loot in an Indian mound surrounded by trees near the present John Lott Monument.

Jackson County

- The Route 64 Bridge Treasure: $40,000 in gold coins hidden a few miles north of Sabula, where the Route 64 bridge crosses the Mississippi River into Wisconsin.
- The River Pirates' Lost Cache: Hidden in a cave near the Mississippi River on the southern side of Bellevue on Route 52.

Lee County

- Black Hawk's Gold: Large amounts of gold coins hidden in the hills near Denmark on Route 103.

Louisa County

- Hoards of the River Pirates: Several such stashes are reported near or at Burris City, on the north side of the Iowa River where it enters the Mississippi.

Monroe County

- Black Hills Lost Treasure: Three prospectors returning from the Black Hills buried $25,000 in gold nuggets sealed in ceramic jars, 1 mile north of the cemetery near Eddyville on the Des Moines River near Route 137.

Page County

- Shenandoah's Lost Cache: $75,000 in gold coins hidden by a merchant near the west bank of the Nishnabotua River, a mile north of Shenandoah on Route 59.

Pottawattamie County

- Jesse James's Gold: The famous outlaw buried $35,000 in gold coins on a farm near Weston, a few miles north of Council Bluffs on Route 191.

Winnshiek County

- The Lost Army Payroll: $7,000 in gold coins buried in the vicinity of Fort Atkinson, near Decorah on Route 9.

Woodbury County

- The River Pirates' Treasures: Several caches are reported in Stone Park, overlooking the Missouri River in Sioux City.

OKLAHOMA

The area that is now Oklahoma has an interesting history. Both the French and Spanish were active here throughout the eighteenth century, and in 1830, the area was designated as Indian Country, and it became the new home of all the people who came here on the Trail of Tears after being evicted from their ancestral homes in the Southeast by the Indian Removal Act of 1830. The name Oklahoma is derived from the Choctaw phrase *okla humma*, meaning "the home of red people." In fact, until 1907, the eastern part of what is now Oklahoma was still officially known as Indian Territory.

Naturally, the treasure lore of Oklahoma is influenced by the Spanish and the French as well as by the Native Americans, but there are also caches stashed by bandits and folk heros from Myra Maybelle Shirley (better known as "Belle Starr") to the Dalton Gang. Jesse and Frank James, along with *their* gang, were particularly active here, and there are a number of reputed hiding places in Oklahoma that contain portions of their ill-gotten gains.

Atoka County

- The Cattleman's Gold of Atoka: In the vicinity of Atoka.

Blaine County

- The Yeager-Black Gang Treasure: In the vicinity of Eagle City.
- The Roman Nose Treasure: Located near Eagle City.

Bryan County

- The Lost Cannon of Fort Washita: In the vicinity of the site of Fort Washita, near Durant.

Caddo County

- The James Gang's Hoard: In the vicinity of Cement.
- The Spanish Treasure of Cement: Located near Cement, Caddo County.

Carter County

- The Gold-Stuffed Cannon in Hickory Creek: At Hickory Creek near Ardmore.

Cherokee County

- The Heirloom Treasure of Chimney Rock: At Chimney Rock, southeast of Tahlequah.

Choctaw County

- The Outlaw Treasure on Boggy Creek: Located on Boggy Creek, near Boswell.

Cimarron County

- The Treasure of Sugar Loaf Peak: In the vicinity of Sugar Loaf Peak or Flag Springs.
- The Treasure of Black Mesa: In the vicinity of Kenton in northwestern Cimarron County.

Coal County

- The Bandit Treasure of Delaware Creek: Located on Delaware Creek.

Comanche County

- Buried Treasure of Cut Throat Gap: At Cut Throat Gap in the Wichita Mountains.

- Dick Estes's Treasure: Located on Panther Creek, about 10 miles north of Cache.

- A Jesse James Treasure: Located on Cache Creek.

- Rattlesnake's Cave of Gold: Hidden in the Wichita Mountains.

- The Buried Spanish Treasure: Located on Mount Scott.

- The Dalton Gang Cache: Hidden in the Keechi Hills near Lawton.

- The Lost Cave With the Iron Door: Located on Elk Mountain in the Wichita Mountains, north of Indiahoma.

- The Lost Platinum Lode of Bat Cave: Hidden in a bat cave in the Slick Hills north of Meers.

- The Mexican Lost Gold of Cache Creek: Located on Cache Creek, south of Geronimo.

- The Outlaw Treasure of Indiahoma: In the vicinity of Indiahoma.

- The Outlaw Treasure of Wildhorse Canyon: Somewhere in Wildhorse Canyon.

- The Spanish Pack Train Treasure at Pecan School: At the old Pecan School, south of Lawton.

- The Spanish Treasure of Hobbs Canyon: Hidden in Hobbs Canyon, west of Meers.

- The Treasures of Fort Sill: Located on or near Fort Sill Military Reservation.

- The Treasure of Seven Springs: At Seven Springs, east of Lawton.

Creek County

- The Treasure of the Dalton Gang: In the vicinity of Mannford.

Delaware County

- The Treasure of Cherokee Lacy Mouse: Hidden in the Spavinaw Hills near Kenwood.

Ellis County

- Diego Parilla's Treasure: Located on Wolf Creek.

Haskell County

- Belle Starr's Treasure: Hidden in the vicinity of Younger's Bend on Canadian River, near Briartown.
- The Lost Standing Rock Mine: At or near Standing Rock.

Hughes County

- The Treasure of the California Emigrants: Located on Fish Creek, near the old Edwards Post, south of Holdenville.

Johnston County

- A Jesse James Cache: Located at or near the place known as "Devil's Den."

Kiowa County

- The Camp Radzimski Treasure: At the site of Camp Radzimski near Snyder.
- The Spanish Treasures in the Wichita Mountains: Hidden throughout the Wichita Mountains, in both Comanche and Kiowa counties.
- The Treasure of Twin Mountain: Hidden on Twin Mountain.
- Bob Herring's Treasure: In the Wichita Mountains near Mangum.
- The Spanish Gold of Devil's Canyon: In Devil's Canyon on Quartz Mountain, near Snyder.
- Hallet's Lost Gold Mine: Somewhere on Rattlesnake Mountain.

Latimer County

- The Chief Blackface Treasure: In the Cherokee Hills near Wilburton.
- Belle Starr's Cache: In "Robber's Cave," north of Wilburton.
- A Hidden Cache of the James Gang: In the San Bois Mountains.

Le Flore County

- A James Gang Cache: Hidden in the Turkey Mountains.
- The Bank Robber's Treasure: Hidden on the Holsum Valley Road.
- The Lost Silver Dollar Cache: Hidden near Summerfield.
- The Lost Spanish Gold Mine on Buzzard Hill: Hidden on Buzzard Hill north of Spiro.

Love County

- The Starling Treasure: In the vicinity of Marietta.
- The Treasure of Sivill's Bend: Located near Sivill's Bend.

Mayes County

- Henry Starr's Buried Treasure: Buried near or between Rose and Pryor.
- The Dalton Gang's Hoard: Cached near Pryor.
- The Meadows Gang Treasure: Located on Pryor Creek near Pryor.
- The Younger Gang's Treasure: Hidden near Pryor.
- The Spanish Treasure of Grand River: Hidden on the banks of the Grand River near Pryor.

McCurtain County

- The Treasure Cave of Pine Knott Crossing: In a cave near the Pine Knott Crossing on Little River, near Valliant.

McIntosh County

- The Chief Yahola Treasure: Located north of Brush Hill and southwest of Checotah.
- Walter Grayson's Treasure: Located on the south branch of the Canadian River near Eufaula.
- The Spanish Treasure at Standing Rock: At Standing Rock on the South Canadian River near Eufaula.

Murray County

- The Treasure of Fort Arbuckle: In the vicinity of the site of Fort Arbuckle, on Mill Creek near Davis.
- The Silver Treasure of Eight Mile Creek: Located on Eight Mile Creek west of Davis.

Muskogee County

- Joe Vann's Treasure: Located on the site of old Vann mansion at Webbers Falls.

Noble County

- The Robber's Treasure at Red Rock: East of Red Rock.

Nowata County

- The Gangster's Treasure: About two miles south of Caney, Kansas near US 75.

Okfuskee County

- The Spanish Gold at Devil's Half Acre: About 8 miles south of Ox-Bow Lake.
- The Buried Gold of the California Miners: Located near Dog Ford on the North Canadian River.

Oklahoma County

- The Spanish Treasure of Lost Creek: Located on Lost Creek near Oklahoma City.

Osage County

- Golden's Wagon Train Treasure: Hidden on Big Caney Creek south of Artillery Mound near Boulanger.
- The Dalton's Hidden Treasure: In the vicinity of Nowata.

Ottawa County

- The Locust Tree Mexican Treasure: East of Miami.
- The Shoemaker's Treasure: At Devil's Promenade, east of Quapaw.

Pawnee County

- The Lost Indian Gold of Twin Mounds: Hidden at Twin Mounds near Jennings.
- The Paymasters Treasure of Twin Mounds: Hidden at Twin Mounds near Jennings.

Payne County

- The Outlaw Treasure of Ingalls: Hidden somewhere in the ghost town of Ingalls, east of Stillwater.

Pontotoc County

- Ben Marshall's Treasure: Hidden near Stonewall.
- The Mission Treasure on the Spanish Trail: In the vicinity of Ada.

Pushmataha County

- The Jewelry Store Robbery Cache: Hidden at the western tip of the Kiamichi Mountains.

- The Glass Jar Treasure: Hidden in the vicinity of the Sulphur Canyon Bridge near Clayton.
- The Lost Indian Mines of the Jack Fork Mountains: Somewhere in the Jack Fork Mountains in western Pushmataha County.
- The Lost Mines and Treasures of the Kiamichi Mountains: Somewhere in the Kiamichi Mountains.
- The Kosoma Train Safe Treasure: Located on Buck Creek near Antlers.
- Zip Wyatt's Lost Gold: Located on Glass Mountain near Fairview.

Rogers County

- The Treasure of Scaley Back Mountain: Located on Scaley Back Mountain near Claremore.
- The Indian Corncrib Treasure: Located on east bank of the Verdigris River at its junction with Bird Creek in southern Rogers County.

Sequoyah County

- Charles "Pretty Boy" Floyd's Treasure: Hidden during his crime spree in the early 1930s near Sallisaw.
- Phillip Ursay's Treasure: Hidden near Sallisaw.
- The Spanish Treasure of Brushy Mountain: Somewhere on Brushy Mountain, which is north of Sallisaw.
- The Treasure of Lee's Creek: Hidden somewhere along Lee's Creek.

Stephens County

- The Spanish Treasure on Mud Creek: Hidden somewhere along Mud Creek.

Tulsa County

- A Dalton Gang Cache: In the vicinity of Sand Springs.
- The Spanish Gold at Tulsa: Hidden near the Arkansas River, west of Tulsa.

Washita County

- The Spanish Gold of Turkey Creek: Hidden somewhere along or adjacent to Turkey Creek, south of Canute.

KANSAS

The plains of the Sunflower State have concealed the treasure of passersby since the sixteenth century. In 1540 and 1541, the Spanish explorer Francisco de Coronado and his expedition surveyed the American Southwest and rode as far as what is now Kansas, where it is said he buried treasure—possibly pirated Indian gold—near the present site of Dodge City.

Since that time, there have been many legends and stories told of other travelers' treasures, left by westward emigrants,'49ers headed to and from California on the likes of the Old Neosho Trail. There are "Bandit" treasures, "Prospector" treasures, and the mysterious "Indian Worshipper" treasure near Osborne.

Cherokee County

- The Treasure of the Neosho Trail: Located on the Old Neosho Trail near Baxter Springs.

Clark County

- The Treasure of Big Basin: Hidden near Ashland.
- Dutch Henry's Treasure: Located on Hackberry Creek.

Douglas County

- The Army Paymaster's Treasure: Hidden near Lawrence.

Edwards County

- The Treasure of Fort Riley: Located at old Fort Riley, southwest of Offerle.
- The '49ers' Wagon Train Treasure: Southwest of Offerle.

Ford County

- The Wells Fargo Treasure of Dodge City: Buried somewhere west of Dodge City.
- Coronado's Treasure: Reportedly buried about 4 miles east of Dodge City.
- The Jesus Martinez Wagon Train Treasure: Somewhere near old Fort Dodge, 5 miles east of Dodge City.

Franklin County

- Ernest De Boissiere's Treasure: At the old De Boissiere silk ranch near Silkville.

Graham County

- The Spaniard's Buried Treasure: In or near Morland.

Montgomery County

- The Bloody Bender's Treasure: Located at or near the Bender Mounds, about 11 miles west of Parsons.

Morton County

- The Bandit Treasure of Elkhart: At Point of Rocks, north-west of Elkhart.

Nemaha County

- The Prospector's Treasure: Located near Seneca, on south bank of the Nemaha River.

Osborne County

- The "Indian Worshipper's" Treasure: Hidden somewhere in the vicinity of Osborne.

Rice County

• The Chavez Treasure: Hidden near Jawis Creek.

Saline County

• The Indian Treasure of Palmer's Cave: At or in Palmer's Cave, near Salina.

Sedgewick County

• The Arkansas River Treasure: At or near the junction of Arkansas and Little Arkansas Rivers, near Wichita.
• Rowdy Joe's Gold: Located on the west bank of the Arkansas River, at or near Wichita.

NEBRASKA

The treasure lore of Nebraska contains many interesting tales, including the story of Neapolis. As every schoolchild knows, Lincoln is now Nebraska's capital. However, this was not always a foregone conclusion. Back before the Cornhusker State entered the Union in 1867, the town of Neapolis, on the south side of the Platte River near Cedar Bluffs, was selected as the state capital. In anticipation of this momentous event, liquor and other merchandise was hidden in the area until suitable buildings could be built. However, the legislature reversed itself and Neapolis was abandoned. It is now a ghost town, and rumor has it that much of the loot still lies buried here.

During the Civil War, the Confederate raiders under William Clarke Quantrill raided the town of Bloomington. Residents buried their valuables before the raid, but many were killed and their caches were not recovered.

As was the case in other Plains states, miners that were headed east from California after the Gold Rush often hid gold and other valuables for later retrieval. One such cache includes a small chest hidden at a place called Point of Rocks, east of Bronco Lake near Alliance. Another is a $70,000-cache buried on the Middle Loup River. Buffalo Bill Cody himself hid $17,000 in gold coins near Scott's Rest Ranch in the vicinity of North Platte, and the lesser-known—but equally colorful— Flyspeck Bill stashed $100,000 in gold and silver coins in several caches near Rushville.

Another intriguing story tells of $100,000 in "Mormon" gold and silver coins that was found in 1946 and reburied on an island in the Platte River near Wood River. A good place to start would be to refer to back editions of newspapers published in and around Hall County in 1946.

Of the fifty states that we contacted for clarification of laws governing treasure hunting, Nebraska was one of eleven that supplied useful information. The office of their attorney general replied by simply referring us to the state's abandoned property statues 69-1316 through 69-1332. Was the Hall County treasure really "abandoned"?

Adams County

- Buried Indian Treasure: In the vicinity of Hastings.

Box Butte County

- The Lost California Gold: A small chest hidden at Point of Rocks, east of Bronco Lake near Alliance on Route 385.

Buffalo County

- The Lost Treasures of Buffalo County: In the general vicinity of Dobytown, Fort Kearny, and Kearny.
- The Bank Robbery Treasure: Hidden near Kearny.

Burt County

- The Ferryman's Caches: The ferryman collected his tolls in empty nail kegs and when full, he buried them near his house on the Missouri River at Decatur.

Cheyenne County

- The Treasure Express Stagecoach Robbery: Four hundred pounds of gold nuggets (worth about $2 million in the 1990s) are said to have been buried on the south banks of the Lodgepole River, 2 miles east of Sidney on Route 30.

Custer County

- The 1921 Kearny Bank Robbery Loot: The robbers buried $40,000 in gold coins and jewelry near the Middle Loup River, 2 miles south of Sargent on Route 183.

Dawes County

- The Lost Cache of Fort Robinson: At Fort Robinson, near Crawford on Route 20.

Dawson County

- Hidden Caches of Rylander: In and around the ghost town of Rylander, 3 miles northwest of Franam on Route 23.
- The Highwayman's Treasure: At or near Lexington.
- The Wiggons Ranch Caches: Numerous caches of gold bullion and coins on buried on this ranch, located between the north side of the North Platte River and Gothenburg on Route 30.

Dodge County

- The Buried Treasure of Neapolis: In the vicinity of Neapolis, a ghost town on the south side of the Platte River, near Cedar Bluffs and 10 miles west of Fremont on Route 30.

Franklin County

- The Treasures of Bloomington: During the Civil War, residents of Bloomington buried their valuables around town before an attack by Quantrill's Raiders.

Garden County

- Garden County's Lost Caches: Rumored to be located throughout the county at such sites as Ash Hollow, Chimney Rock, and Courthouse Rock, all on the north side of the North Platte River west of Lewellen on US 26.

Hall County

- The Mormon Treasure: $100,000 in gold and silver coins found in 1946 and reburied on an island in the Platte River near Wood River on Route 11.

Hooker County

- The Cattle Thieves Loot: $25,000 in gold coins buried near Mullen on Route 2, Hooker County.

Knox County

- Devil's Nest: Along the Missouri River, about 5 miles north of Linky on Route 12, at or near Devil's Nest favorite hide-out for nineteenth-century outlaws.

Lincoln County

- Buffalo Bill Cody's Cache: $17,000 in gold coins hidden near Scott's Rest Ranch near North Platte on Route 70.

Morrill County

- The Mud Springs Cache: $20,000 in gold coins was buried near Mud Springs on the north side of the North Platte River about a mile north of Broadwater on Route 92.

Otoe County

- The Jesse James Gang Loot: Hidden on the Catron-Miyoshi Fruit Farm, 3 miles southeast of Nebraska City on Route 35.

Rock County

- The Pony Boys' Cache: Stagecoach loot buried near a stream in the vicinity of Bassett on US 20.

Sarpy County

- The Bootlegger's Cache: In the 1930s he buried $10,000 in gold coins and $25,000 in currency near the Missouri River in the vicinity of the Plattsmouth Bridge off Route 34.

Scott's Bluff

- The Scott's Bluff Treasure: Cached on the south bank of the South Platte River near Scott's Bluff on Route 92.

Sheridan County

- Flyspeck Bill's Treasure: $100,000 of gold and silver coins buried in several caches near Rushville on Route 20.

Thomas County

- The California Miner's Lost Gold: $70,000 in gold is buried along the banks of the Middle Loup River, 2 miles from Seneca on Route 2.

Thurston County

- The Robber's Cave Bandit Treasures: In the vicinity of Macy at the place known as Robber's Cave.

SOUTH DAKOTA

The areas that are now North and South Dakota were designated in 1861 as "Dakota Territory," and they entered the Union twenty-eight years later as two separate states. In that relatively short time, the region had been transformed from vast open prairie populated mainly by nomadic Native Americans (principally the Souian-speaking Dakota people) to an agricultural region populated primarily by white immigrants from the East.

The single most important event in Dakota history in this period—and in the area's treasure lore—was the 1874 discovery of gold in the Black Hills region of what is now South Dakota. This initiated a massive gold rush into an area previously ceded to Native Americans, angering them and leading in turn to a series of events that included Colonel George Armstrong Custer's waterloo—the Battle of the Little Bighorn (1876)—in neighboring Montana and the subsequent forced end to the Native peoples' nomadic way of life.

For white settlers, the gold rush created the burst of economic development called the "Dakota Boom," which lasted into the 1880s and changed the region economically forever. With the Dakota Boom came the "boom towns" in the Black Hills such as Deadwood, Lead, Sturgis, and Rapid City, which today are popular tourist destinations, and are steeped in the treasure lore of the region and the state. The Black Hills would be an ideal place to begin a South Dakota treasure tour, and a word or two dropped at the right saloon, library, or museum would start to flesh out the information given below.

Of the fifty states that we contacted for clarification of laws governing treasure hunting, South Dakota was one of eleven that supplied useful information. The office of their attorney general replied by referring us to South Dakota Codified Laws Chapter 1–20, under which anyone conducting an "archeological exploration" is required to obtain a permit from the State Historical Society. For more information, contact

Office of the Attorney General
500 East Capitol
Pierre, South Dakota 57501-5070

Butte County

- The New Yorker's Treasure: In the vicinity of Belle Fourche.

Codington County

- The Buried Treasure of Long Lake: At or near Long Lake.

Custer County (Black Hills)

- Desseri's Lost Gold Hoard: In the Black Hills near Galena.
- Ezra Kind's Lost Gold: Somewhere in the Black Hills.
- Joseph Metz's Treasure: At Red Canyon near Custer.
- Lame Johnnie's Buried Treasure: In the vicinity of Horse Thief Lake.
- Mexican Ed Chanchez's Treasure: Located on Dirty Woman Creek, near the old town of Grindstone.
- The Stagecoach Holdup Treasure: Hidden on French Creek near Fairburn.
- Toussaint Kensler's Treasure: Somewhere on the south bank of the Cheyenne River.

Fall River County (Black Hills)

- The Treasure of Hat Creek: Somewhere near Rumford in the vicinity of Hat Creek.
- The Sidney Stagecoach Treasure: Hidden along Hat Creek near Ardmore.
- The Lost Gold of Red Canyon: In Red Canyon south of Edgemont.

Hughes County

- The Buried Treasure of the Missouri: In the vicinity of Pierre.
- The Three Sisters' Treasure: In or near Pierre.

Lawrence County (Black Hills)

- Archie McLaughlin's Treasure: Hidden near Deadwood.
- General Custer's Lost Rifle Cache: South of Mystic, in Pennington County or near Nemo in Lawrence County.

- The Buried Treasure of Gordon Stockade: Located on French Creek near the site of the Gordon Stockade.
- The Four Directions Treasure: Hidden near Deadwood.
- The Holy Terror Mine Treasure: Located near Deadwood, Lawrence County.
- The Lost Cabin Gold Mine: Located in the vicinity of White-wood.
- The Lost Raspberry Mine: Somewhere near Central City.
- The Seven Missourians' Lost Mine and Treasure: Said to be located on Lookout Mountain.

Marshall County

- Grey Foot's Treasure: At east end of Long Lake near Lake City.

Minnehaha County

- Jesse James's Treasure: At Jesse James Cave near Garretson.

Pennington County (Black Hills)

- General Custer's Lost Rifle Cache: South of Mystic, in Pennington County or near Nemo in Lawrence County.
- The Burnt Ranch Treasure: In the vicinity of Redfern.
- Shafter's Lost Tub of Gold: Said to have been hidden near Hill City.
- The Scruton Brothers' Lost Mine: Located on or near Scruton Mountain.
- The Treasure of Spring Creek: Located on Spring Creek near Rockerville.
- Proteau's Gold: Somewhere on or near Rapid Creek.
- Lame Johnny's Treasure: In the vicinity of Horse Thief Lake.
- Gratz Duke's Lost Mine and Treasure: In the vicinity of Castle Creek.

- The Lost Treasure of the Four Crosses: Located on or near Castle Creek.
- The Lost Treasure of the Limestone Cave: In or in the vicinity of Limestone Cave near Rapid City.
- The Canyon Springs Stage Station Gold: In the vicinity of a former stagecoach station at the junction of Prairie Creek and Rapid Creek near Rapid City.

NORTH DAKOTA

In terms of treasure lore, South Dakota was blessed with Black Hills gold mines, many of which are still out there and still potentially profitable. In North Dakota, as with the other Plains states farther south and east, most of the treasures that lie awaiting discovery were lost or hidden there by their rightful or wrongful owners.

This has given us some interesting tales that date back into the eighteenth century, when British fur traders working for the Hudson's Bay Company prowled the landscape. For example, during an Indian attack, a Hudson's Bay Company paymaster buried $40,000 on Big Butte near Lignite. A trader, possibly affiliated with the "Bay" Company, buried $55,000 in gold and silver coins on the east bank of the Missouri River at the mouth of Burnt Creek between Mandan and Bismarck.

Lost gold is always a staple in tales of what might have been. There is said to be $100,000 of California gold buried near the Knife and Missouri rivers, a mile north of Stanton, and another hoard of $200,000 in gold bullion hidden in Stanton. During an Indian attack, another $90,000 in Montana gold was buried along the Missouri River, about a mile east of Fort Clark. In addition to "lost" gold, there are always tales of stolen gold, such as the several chests of gold bullion hidden by bank robbers during the 1880s in what is now the ghost town of Pleasant Lake, or the $100,000 in pilfered plunder hidden in 1893 near Belcourt in the foothills of the Turtle Mountains.

The most tantalizing stories in treasure lore involve hoards that were lost, then found, then lost again. Such is the case of what is reported as a "Lost Mountain of Gold." As the story goes, it is a mine somewhere near Belfield, that was originally discovered in 1864, and later lost.

Of the fifty states that we contacted for clarification of laws governing treasure hunting, North Dakota was one of eleven that supplied useful information. Unlike other states, North Dakota has not adopted a uniform treasure trove law, but some of the general statutes may be applicable. North Dakota Century Code (NDCC), section 55-02-07, provides that "any historical . . . artifact or site that is found or located upon any land owned by the state of North Dakota or its political subdivisions . . . which is significant in understanding and interpreting the history and prehistory of the state, shall not be destroyed, defaced, altered, removed, or otherwise disposed of in any manner without the approval of the state historical board."

NDCC (section 55-03-01.1) also prohibits individuals from investigating or excavating cultural resources on land owned by the state of North Dakota and in the excavation of cultural resources on private land without a permit from the Superintendent of the State Historical Society of North Dakota. North Dakota Administrative Code (chapter 40-02-02) sets forth the permit application criteria. This code section provides that "all activities performed under a permit must be conducted by or under the direct supervision of a professionally qualified individual."

NDCC section 23-06-27 provides for the protection of human burial sites. NDCC section 23-06-27(3) provides that "[a] person is guilty of a felony who, without authority of law, willfully . . . disturbs a human burial site, human remains, or burial goods found in or on any land, or attempts to do the same, or incites or procures the same to be done."

Benson County

- The Pleasant Lake Cache: Several chests of gold bullion hidden during the 1880s in the ghost town of Pleasant Lake, 45 miles northwest of Devil's Lake on US 2.

Burke County

- The Hudson's Bay Company Payroll: $40,000 buried on Big Butte near Lignite on Route 52.

Burleigh County

- The Buried Treasure of Burnt Creek: In the vicinity of Burnt Creek near Bismarck.

McLean County

- The Lost Miner's Buried Gold: Hidden near Garrison.

Mercer County

- Lost California Gold: $100,000 of California gold was buried near the Knife and Missouri Rivers 1 mile north of Stanton.
- Lost California Gold: $200,000 in gold bullion hidden in Stanton, along the banks of the Missouri River.

Morton County

- The Trader's Lost Hoard: $55,000 in gold and silver coins buried on the east bank of the Missouri River between Mandan and Bismarck, at the mouth of Burnt Creek, a quarter mile north of the railroad bridge.

Oliver County

- The Lost Montana Gold: $90,000 in Montana gold buried along the Missouri River about a mile east of Fort Clark on Route 200A.

Rolette County

- The Buried Treasure of Turtle Mountain: In the Turtle Mountains near Dunseith.
- The Lost Bank Robbery Cache: $100,000 hidden in 1893 near Belcourt in the foothills of the Turtle Mountains.

Slope County

- The Lost Treasure of Chalky Butte: Located on or in the vicinity of Chalky Butte near Amidon.
- The Sunset Butte Army Payroll: Stolen gold coins buried on Sunset Butte, 10 miles northwest of Amidon on Route 85.

Stark County (near or across its border with Billings County)

- Dr. Dibb's Lost Mine: In the vicinity of Belfield.
- The Lost Mountain of Gold: A mine somewhere near Belfield on what is now Interstate 94.

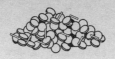

10

THE TREASURES OF TEXAS

With contemporary treasure in the form of oil, cattle, and high technology, Texas is a rich and powerful state, as it has been throughout its long and colorful history. With the exception of California, the Lone Star State also has more hidden, lost, or buried treasures than any other state. In the sixteenth and seventeenth centuries, when Spain claimed Texas, the conquistadors transported vast quantities of Mexican and Native American gold across the region. In the eighteenth century, pirates such as the legendary Jean Lafitte cruised the islands and bayous of the Gulf Coast.

Jean Lafitte's name is indeed commonly heard along the Gulf Coast from Beaumont to Padre Island. With someone so notorious, it is hard to sort fact from fiction, as almost any alleged "pirate" treasure would be linked in legend to the most famous of the region's buccaneers. We do know that Lafitte maintained an operational headquarters on Campeche, or "Campeachy" Island, which is now Galveston Island. Today, one can pick up the thread of a Lafitte buried treasure story in almost any county in south or southeast Texas. The locales most frequently mentioned are Brazoria County, Calhoun County, Chambers County, Galveston County, Harris County, Harrison County, Jackson County, Jefferson County, Kleberg County, and Nueces County.

In the nineteenth century, Texas won its independence from Mexico, existed for a decade as a republic, and joined the

Union in 1845. It was also during the nineteenth century that Texas became a cattle-producing region of unprecedented proportion, and with the cows came the cowboys, the cowtowns, and the wildness that characterizes our image of the "Wild West." Names like Jesse James and Sam Bass figure prominently in the history and treasure lore of the Lone Star State. As with Jean Lafitte and pirate loot, many of the caches rumored to have been secreted by Jesse James may well have been hidden by a desperado of lesser historical stature.

Another name well known to treasure hunters is that of Karl Steinheimer, a former German merchant seaman and sometimes pirate. Fortunes, especially in those of pirates, have a tendency to come and go, and Steinheimer lost his. It was in about 1830 that Steinheimer left the sea and went deep into the mountains of Mexico to prospect for gold. He struck it rich and was heading north with a reported ten mule loads of gold and silver when Texas declared its independence from Mexico. Being neither a Texan nor a Mexican, Steinheimer was caught in the middle.

Realizing that his treasure was probable plunder for either faction, Steinheimer decided that he'd best bury it until things cooled down a bit. He was about fifty miles north of Austin when he started looking for a place in the rolling hills that would form a cache that would be easy to remember. Ultimately, he buried the entire multimillion dollar treasure under a tree (some stories say it was an oak) near (some stories say a few miles south of) Temple near where the Leon and Lampasas Rivers join to form the Little River. To make sure that he'd know *which* tree, he pounded a brass spike into the trunk.

A few hours later, the German and two travelling companions were attacked by Indians. Steinheimer was the only survivor, but he was badly wounded, and by the time he was found a few days later by a group of Texans, he was near death. Before he died, however, Karl Steinheimer managed to write a letter to a former girlfriend in St. Louis describing his cache.

Because of the war in Texas and various other factors, it was several years before friends of the Missouri woman reached Texas. They found the place where the Leon and Lampasas Rivers join to form the Little River, but they never found the tree. As legend has it, no one has ever found the tree.

The end of the Civil War in 1865 left us with many rumors of stolen, lost and/or hidden Confederate gold throughout the former Confederacy states, such as Texas and adjacent states. At the same time, geopolitical intrigue was adding to the treasure lore south of the border in Mexico. France's star-struck, self-styled emperor Napoleon III chose the moment of America's preoccupation with the Civil War to seize control of Mexico's government. He declared Mexico to be an "Empire" allied with France and in 1864, he installed an Austrian Archduke named Ferdinand Maximilian to be "Emperor" of Mexico. The farce lasted only a few years and the Austrian was overthrown and killed by his Mexican "subjects" in 1867.

During his few years in power, however, Maximilian demonstrated a voracious appetite for riches, and incredible quantities of gold and silver flowed into his palace from Mexico's rich and profitable mines. By 1866, however, the greedy Max sensed that his time was running out and he decided to send his treasure home by way of Texas. Fifteen wagon loads of gold and silver coins as well as jewels and other treasure disguised as barrels of flour were sent north and successfully crossed into West Texas near El Paso.

En route to San Antonio, the Austrians guarding the wagon train met a group of Confederate troops escaping to Mexico. The six Confederates reported Indians and bandits on the road ahead, and were hired to help protect the "flour." A few days later, at Castle Gap west of the Pecos River, the Confederates discovered that it was not flour, and killed the Austrians. Having loaded what they could in their saddlebags, the Confederates buried the rest along with the bodies of the Austrians, and burned the wagons atop the spot.

One Confederate, suspected of betrayal, was shot and left for dead, but the other five were later killed by Indians and the first man survived. He was rescued by a group of cattle rustlers, and was captured when they were arrested. Ultimately, he died of the wounds he suffered when he was shot, but he managed to tell the doctor who cared for him about Maximilian's treasure. It was several years before the doctor was able to reach Castle Gap, and by that time, no trace of the burned wagons could be found.

This amazing tale is only one of many stories of lost wagon

trains or mule trains of Spanish or Mexican origin lost or hidden somewhere in Texas.

Of the fifty states that we contacted for clarification of laws governing treasure hunting, Texas was one of eleven that supplied useful information. We contacted the Texas Attorney General's Office for information and were referred to the Texas Historical Commission, the agency responsible for administration of antiquities in this state. The Historical Commission has published many informative booklets and circulars dealing with antiquities conservation and legislation in Texas and these may be obtained by writing to

Department of Antiquities Protection
Texas Historical Commission
PO Box 12276
Austin, Texas 78711
(512) 463-6096

In general, there are two fundamental requirements that must be met prior to treasure hunting on any state-owned lands in Texas: (1) securing an antiquities permit from the Texas Historical Commission, and (2) securing a permit or other access agreement from the agency that controls the land.

In general, the Historical Commission considers unregulated treasure hunting as inimical to the best interest and stated policy of the state, because of its potential for adversely affecting significant cultural resource sites. Such activities on the 20.5 million acres of state-owned lands are covered by the declared policy of the state, found in the Texas Natural Resources Code, Chapter 191, whose purpose is to protect and preserve significant cultural resources.

Relevant excerpts from Chapter 191 include the following.

It is the public policy and in the public interest of the state of Texas to locate, protect, and preserve all sites, objects, buildings, pre-twentieth-century shipwrecks, and locations of historical, archeological, educational, or scientific interest, including but not limited to, prehistoric and historical American Indian or aboriginal camp sites, dwellings and habitation sites, archeological sites of every character, treasure embedded in the earth, sunken or abandoned ships and wrecks of the sea or any part of their contents, maps, records, documents, books, artifacts, and implements of culture in any way related to the inhabitants, pre-history, history, natural history, government, or culture in, on, or under any of the land in the State of Texas, including the tidelands, submerged land, and the bed of the sea within the jurisdiction of the State of Texas (Chapter 191.002, Texas Natural Resources Code, Vernon Supp. 1994).

Sunken or abandoned pre-twentieth-century ships and wrecks of the sea, and in any part or the contents of them, and all treasure embedded in the earth, located in, on or under the surface of land belonging to the State of Texas, including its tidelands, submerged land, and the beds of its rivers and the sea within jurisdiction of the State of Texas, are declared to be state archeological landmarks and are eligible for designation (Chapter 191.091, Texas Natural Resources Code, Vernon Supp. 1994).

Sites, objects, buildings, artifacts, implements, and locations of historical, archeological, scientific, or educational interest, including those pertaining to prehistoric and historical American Indians or aboriginal campsites, dwellings and habitation sites, their artifacts and implements of culture, as well as archeological sites of every character that are located in, on, or under the surface of any land belonging to the State of Texas or to any county, city, or political subdivision of the state are state archeological landmarks and are eligible for designation [Chapter 191.092(a), Texas Natural Resources Code, Vernon Supp. 1994].

Angelina County

- The Robbers' Treasure of Biloxi Creek: Located on or near Biloxi Creek near Lufkin.
- The Treasure Cannon of the Neches River: Located on the Neches River at or near Diboll.

Aransas County

- The Treasure of St. Joseph's Island: Located on San Jose (St. Joseph's) Island offshore from Rockport and north of Corpus Christi.

Archer County

- The Nugget Hole Treasure: Somewhere south of Lake Wichita.

Armstrong County

- John Casner's Treasure: In or near Palo Duo Canyon.

Bandera County

- The Spanish Treasure in Olmos Pass: At or near Olmos Pass.

Bastrop County

- The Spanish Treasure of Bastrop County: Along the Colorado River.

Bee County

- The Kidnapped Man's Treasure: Located on west side of the Old Brownsville Road south of Beeville.

Bell County

- San Miguel Aguayo's Lost Mine: In the vicinity of the Lampasas River between Belton and Salado.
- Karl Steinheimer's Treasure: Ten mule loads of gold and silver buried under a tree (some stories say it was an oak) near (some stories say a few miles south of) Temple (or possible Belton) near where the Leon and Lampasas Rivers join to form the Little River. (The tree has or had a brass spike in the trunk.)
- The Lost Temple of La Grange: Northeast of Temple.
- The Mexican Treasure on the Lampasas River: Located on the Lampasas River near Belton.
- The Golden Bull Cave Treasure: In a cave along Salado Creek near Salado.

- The Lost Indian Gold of Belton: About 18 miles southwest of Belton.

Bexar County

- The Alamo Tunnel Treasure: Hidden in a tunnel under the streets of San Antonio that has one of its ends under or near the Alamo between East Commerce Street and East Houston Street.
- The Illinois Train Robber's Treasure: Hidden near San Antonio.

Brazoria County

- Holt's Rose Garden Treasure: In or near Alvin.
- The Pirate Ship Treasure: Located near the mouth of the San Bernard River.

Brewster County

- Benito Ordones' Lost Gold Bars: At or near Paisano Pass near the Brewster-Presidio county line.
- Sam Bass Caches: Located on Packsaddle Mountain and/or Sue Peak.
- The Humphrey Treasure: In the vicinity of Boquillas.
- The Lost Gold of the Chisos Mountains: In or near the Chisos Mountains in Big Bend National Park.
- The Lost Mine on Horse Mountain: Located on Horse Mountain.
- The Lost Mine of Gold Peak: Located on Gold Peak near Telingua.
- The Treasure of Death Hill: At Death Hill near Emory Peak.
- The Sanderson Train Robbery Treasure: In the vicinity of Sure Peak.
- Gideon Gaines's Treasure: In the Chisos Mountains of Big Bend National Park.
- The Ranger's Treasure: Somewhere on Packsaddle Mountain.

- The Phantom Lost Mine: In the Phantom Mountains.
- Don Luiz Terrazas's Treasure: In the Rosillos Mountains north of Big Bend National Park.
- Seminole Bill Kelly's Lost Ledge: A large deposit of very rich gold ore in or near Reagan Canyon (some stories say the *head* of the canyon named for the Reagan brothers who ranched here in the 1880s), possibly concealed by a landslide occurring before 1890.
- The Treasure of Pine Canyon: In or near Pine Canyon.
- The Engineer's Lost Ledge: At or near Paisano Pass near Alpine.

Burleson County

- Colonel Frisbee's Treasure: Hidden near Caldwell.

Burnet County

- The Nugget Cave Treasure: In a cave near Burnet.
- Samuel McFarland's Lost Silver Mine: In the vicinity of Burnet.
- The Church Treasure of Longhorn Cavern: Somewhere in Longhorn Cavern near Burnet.

Caldwell County

- Pedro Gomez's Lost Silver: In the vicinity of Lockhart.
- The Mexican Silver Bar Mine: Somewhere near Lockhart, possibly related to the preceding entry.
- The Lost Silver of Clear Fork Creek: South of Lockhart.

Calhoun County

- The Mexican Treasure of Tiger Lake: In or near the old Mexican settlement of Tiger Lake.

Callahan County

- The Flying H Ranch Treasure: Located somewhere on the Flying H Bar Ranch near Clyde.

Cameron County

- John Singer's Treasure (aka the Singer Family's Buried Treasure): Located on Padre Island, reachable by the bridge between the towns of Port Isabel and South Padre Island in Cameron County (although the north end of the island is in Willacy County).
- The Prospector's Treasure of Padre Island: Located on Padre Island (*see* previous entry).
- Wailing Wayne's Treasure: Located on Padre Island, near Point Isabel.
- The Buried Treasure of Palo Alto: Somewhere in the Brownsville area.
- The Mexican Treasure of Resaca De La Palma: At or near the Battle site of Resaca de la Palma near Brownsville.
- The Treasure of Lago de los Pajores: At or near Lago de los Pajores near Brownsville.

Coleman County

- The Spanish Treasure of MMM Rock: In the Santa Anna Mountains.
- Chief Satanta's Wagon Train Treasure: In the Santa Anna Mountains.

Cooke County

- The Cross Timbers Treasure: At Cross Timbers between Gainesville and Burns.
- The Old Spanish Fort Treasure: At or near the Cooke-Montague County line.
- The Spanish Treasure of Walnut Bend: Along the Red River near Dexter.

Culbertson County

- The Treasure of Juniper Cave: At Juniper Cave, northeast of Guadalupe Peak (the highest point in Texas) in the Guadalupe Mountains National Park.
- William "Ben" Sublett's Lost Mine: In the vicinity of Culbertson or Hudspeth County near Guadalupe Peak in the Guadalupe Mountains National Park.
- The Old Spanish Gold Mine: Below Signal Peak, at head of Shirt Tail Canyon in the Guadalupe Mountains.

Dallas County

- Jesse James Treasures: Reported to have been hidden in and around Dallas.
- Sam Bass Treasures: Reported to have been hidden in and around Dallas.

Dickens County

- The Croton Breaks Lost Silver Mine: In or near the Croton Breaks, near Gilpin.

Dimmit County

- The Lost Estambel Hill Treasure: At or near Estambel Hill.
- The Treasure of Brand Rock Hole: At or near the Brand Rock Hole, along Pena Creek.

Duval County

- The Treasure of Cerro Del Rico: In the vicinity of San Diego.
- The Mexican Silver of Realitos: South of Realitos.

Eastland County

- Turger's Bank Robbery Treasure: About 8 miles south of Cisco on the south fork of the Leon River.

El Paso County

- Don Nicasio's Treasure: Located on Franklin Mountain near El Paso.
- The Padre's Lost Mine and Treasure on Franklin Mountain: Located on Franklin Mountain near El Paso.
- James B. Leach's Treasure: Hidden somewhere in the vicinity of El Paso.
- Pancho Villa's Treasure: Said to have been hidden somewhere in the Franklin Mountains.

Galveston County

- Ham Washington's Treasure: Somewhere on Galveston Island.
- The Treasure of Pelican Island: Somewhere on Pelican Island.
- The Muscatee Men Treasure: Located near Bolivar Point north of Galveston.

Grayson County

- Jesse James' Lost Hoard: In the vicinity of Pottsboro.

Hansford County

- The Treasure of the Rifle Pits: Located on Palo Duro Creek in northeastern Hansford County.

Harris County

- The Lost Gold Bullion of Dead Man's Lake: At or near Dead Man's Lake, in the vicinity of Humble.
- The Treasure of Piney Point: Located near Piney Point west of Houston.
- Enoch Bronson's Treasure: West of La Porte.

Harrison County

- The Confederate Walking Cane Treasure: In the vicinity of Marshall.
- The Caddo Indian Lost Lead Mine: In the vicinity of Scottsville.
- The Hendricks Lake Treasure: At or near Hendricks Lake near Tatum.

Haskell County

- The Lost Copper Mine of Double Mountain Fork: At Double Mountain Fork west of Jud.

Hays County

- L. J. Daily's Lost Silver Ledge: At or near Shelton Holler, southwest of Wimberly.
- The Spanish Train Treasure on Onion Creek: In the vicinity of Onion Creek.

Hill County

- The Hooker Farm Treasure: Located on the old Hooker farm near Abbot.

Howard County

- The Treasure of Signal Mountain: Somewhere on Signal Mountain.

Hudspeth County

- The Gold Bullion of Apache Canyon: At Apache Canyon on Cerro Diablo Mountain in northern Hudspeth County.
- The Lost Royal Crown Jewels: In the vicinity of Pine Spring in the Guadalupe Mountains.
- Jacob Snively's Lost Mine: Located on Eagle Mountain.

- Quick Killer's Lost Mine: Somewhere in the Guadalupe Mountains.
- William "Ben" Sublett's Lost Mine: In Culbertson or Hudspeth County near Guadalupe Peak in the Guadalupe Mountains National Park.
- Policarpo Gonzales's Lost Mine: In the Guadalupe Mountains.

Jasper County

- The Mexican Treasure of Jasper County: North of Neches River near the border between Jasper and Angelina counties.
- The Gold-Filled Cannons: At Bone's Ferry on the Neches River in Jasper County (also at the Big Sandy River in Lavaca County).

Jeff Davis County

- Red Curley's Buried Bandit Treasure: In the Davis Mountains.
- The Ebony Cross Treasure: In the vicinity of Valentine.
- The Lost Bandit Treasure of Seminole Hill: In the Davis Mountains.
- The Lost Treasure of Barrel Springs: At or near Barrel Springs.
- The Lost Treasure of Deadman Springs: At or near Deadman Springs.
- The Treasures of El Muerto Springs: At El Muerto Springs in the Davis Mountains.

Jefferson County

- Neil McGaffey's Treasure: At Shell Ridge near Sabine Pass.
- The Pirate Treasure of Sabine Pass: At or near Sabine Pass.

Jim Wells County

- The Treasure of Casa Blanca: In the vicinity of old Spanish fort near Casa Blanca in northeastern Jim Wells County.

Kenedy County

- The Carreta Cart Trail Treasure: Located near an old cart trail in the vicinity of Armstrong.

Kerr County

- Antonio Lopez's Treasure: At or near Elm Pass, south of Center Point.

Kimble County

- The Indian Treasure of the Cedar Breaks: At the Cedar Breaks on the Old Lechuza Ranch.
- Chief Yellow Wolf's Lost Silver Mine: In the vicinity of Junction, in northeastern Kimble County.

King County

- Thomas Longest's Lost Lead Mine: In the vicinity of North Craton Creek.

Kinney County

- Santiago's Cave Treasure: In the vicinity of Spofford.
- The Treasure of the Anna Cache Mountains: Somewhere in the Anna Cache Mountains.

Kleberg County

- The Treasure of Kingsville: In the vicinity of Kingsville.

Lamar County

- The Mexican Pack Train Treasure at Lake Palmer: At or near Palmer Lake.

Lampasas County

- The Lost Treasure of Lometa: In the vicinity of Lometa.
- The Comanche Treasure of Adamsville: Located near Adamsville.

LaSalle County

- Musgraves's Gold Coin Treasure: In the vicinity of Cotulla.
- The Lost Treasure of Sheep Pen Rocks: At or near Sheep Pen Rocks.
- The Mexican Revolution Treasure: Buried somewhere between Cotulla and Artesia Wells.
- The Mexican Treasure of Estambel Hill: At or near Rancho de Los Olmos in southeastern La Salle County.
- The Spanish Treasure of Seven Rock Hill: At or near the place called Seven Rock Hill near Fowlerton.

Lavaca County

- Frank Vanlitsen's Lost Gold: In the vicinity of Hallettsville.
- The Gold-Filled Cannons: Located on the Big Sandy River (also at Bone's Ferry on the Neches River in Jasper County).
- The Lost Dutchman Mine: In the vicinity of Hallettsville. (As with the legendary Lost Dutchman Mine in Arizona, the term "Dutchman" probably implies a German, or "Deutsche" man.)

Lee County

- James Goacher's Lost Lead Mine: Located on Rabb Creek near Giddings.

Leon County

- The Mexican Caravan Treasure: Located on the old road between Navasota and the Trinity River.

Liberty County

- The Priest's Treasure at Big Thicket: At Big Thicket near Kenefick.
- The Spanish Mule Train Treasure on Bowie Creek: Located on Bowie Creek near Dayton.

Limestone County

- Marion Graham's Treasure: In the vicinity of Kosse in southwestern Limestone County.
- John Hightower's Treasure: Somewhere near Kosse.

Live Oak County

- A Jesse James's Cache: At or near Three Rivers.
- The Paso Valeno Buried Treasure: Somewhere near Live Oak.
- The Buried Treasure of Round Lake: At or near Round Lake.
- The Padre's Treasure of Lagarto: In the vicinity of Lagarto in the southeastern corner of Live Oak County.
- The Treasure of Fort Ramirez: At site of old Fort Ramirez in the southeastern corner of Live Oak.
- The Treasure of Puente De Piedra: Located on the west bank of the Nueces River, near the old Laredo-Goliad road which roughly parallels US 59.
- The Treasure of San Casimiro: In the vicinity of Three Rivers.
- The Treasure of Fort Merrill: Located near old Fort Merrill.

Llano County

- Mathews's Lost Mine: Located on the Llano River near Castell.
- Sam Bass's Treasures: In the vicinity of Costulla.
- The Lost Blanco Mine: Located on Packsaddle Mountain.
- The Blind Ranger's Icicle Silver Cave: Somewhere west of the Colorado River. (The term "Icicle" implies the presence of stalactites in this cave.)
- The Buried Treasure of Longhorn Cavern: At Longhorn Cavern near Llano.
- The Texas Ranger's Lost Pick Placer: In the vicinity of Oxford, the site may be identified by finding the pick.
- William Nard's Lost Lead Mine: Located on the point of Cedar Mountain.

McCulloch County

- The Lost Silver Mine of Calf Creek: In the vicinity of the town of Calf Creek in southwestern McCulloch County.

McLennan County

- The Treasure of the Tonkawa Indians: In the vicinity of Crawford.

McMullen County

- Dan Dunham's Lost Treasure: Somewhere in McMullen County.
- The Lost Gold of the San Caja Mountains: In the San Caja Mountains and/or the Little San Caja Mountains.
- The Lost Treasure of Laredo Crossing: Located near Laredo Crossing.
- The Spanish Mule Train Treasure of the Las Chuzas Mountains: Between the Las Chuzas Mountains and the Nueces River.

- The Indian Lost Mine of the Las Chuzas Mountains: In the Las Chuzas Mountains.
- The Treasure of the Rock Pens: In the vicinity of Tilden.
- The Mexican Treasure of Alta Loma Mountain: Located on east slope of Alta Loma Mountain in south central Mc-Mullen County.

Martin County

- The Train Robber's Treasure at Stanton: Reported to have been hidden on the southwestern edge of Stanton.

Mason County

- Reese Butler's Lost Mine: Located near confluence of Threadgill Creek and the Llano River.
- John Gamel's Treasure: Located on the old Gamel homestead.

Maverick County

- Jesse James's Hidden Treasure: At or near Eagle Pass.
- The McNeal Brothers' Treasure: North of Eagle Pass.
- The Newton Brothers' Robbery Cache: In the vicinity of Eagle Pass.

Menard County

- James Bowie's Mine (aka The San Saba Lost Mine, Lost Las Amarillas, Lost Almagres Mine, and Lost La Mina de las Iguanas): In the vicinity of the San Saba River.
- Jack Wilkerson's Treasure: Located on the old Wilkerson Ranch near Hext.
- The Indian Lost Silver of Lost Moros Creek: Located on Lost Moros Creek.

Mills County

- The Treasure of the San Saba Banking Mission: Located near Epley Spring, northeast of Goldthwaite.
- The Spanish Treasure of Epley Spring: In the hills west of Epley Spring near Goldthwaite.
- Sam Williams's Treasure: Located near the old Pendleton Ranch west of Mullin.
- Mose Jackson's Treasure: Hidden near Rattler.

Montague County

- The Old Spanish Fort Treasure: At a former Spanish installation on or near the present Cooke-Montague County line. (A variation on the story suggests that the ''fort'' is about 1 mile south of Red River, and perhaps both indications are correct.)
- The Mexican Treasure at Illinois Bend: In the vicinity of Illinois Bend on the Red River.

Navarro County

- The Outlaw Treasure on Wolf Creek: Probably hidden on or near Wolf Creek.

Nueces County

- Fred Oppercoffer's Treasure: Hidden on Shamrock Island.
- Jacob Zieglar's Treasure: Said to be hidden on north Beach at Corpus Christi.
- John Fogg's Treasure: Hidden in or around Corpus Christi.
- Mayberry B. ''Mustang'' Gray's Treasure: At Rock Crossing on the west side of the Nueces River.
- Santa Ana's Treasure: Located near Rock Crossing on the Nueces River.
- Pirate's Treasure: Various sites near the mouth of the Nueces River around Corpus Christi are said to have been popular with Jean Lafitte and others.

- Sourdough Lassiter's Treasure: Said to be hidden somewhere on the shore of Lake Corpus Christi.
- The Lost Treasure of the Casa Blanca: In the Nueces River area above Corpus Christi in San Patricio or Nueces County.
- The Mexican Treasure of the Riverside Ranch: Located on or near the Riverside Ranch.
- The Money Hill Treasure: At Money Hill on Mustang Island.
- The Treasure of San Pedro: In the vicinity of San Pedro.
- The Treasure on Oso Creek: Located on Oso (Bear) Creek near Corpus Christi.
- Tontino's Buried Treasure Chest: Hidden somewhere in or around Corpus Christi.

Oldham County

- Chilton Graves's Treasure: Buried in the vicinity of Tascosa.

Parker County

- A Sam Bass Treasure: Hidden near Springtown.
- The Bandit Treasure of Weatherford: Hidden at or near the Old Curtis diggings near Weatherford.

Pecos County

- Montezuma's Treasure: The legendary Montezuma Treasure has been placed in Pecos County, Reeves County, Stonewall County, and Williamson County, although the legend is hazy on specific location(s).
- The Seven-Mile Mountain Cache: At Seven-Mile Mountain east of Fort Stockton.

Polk County

- The Lost Gold of Bad Luck Creek: Located near Bad Luck Creek.

Presidio County

- The Hidden Gold in the Chinati Mountains: Located on Cibola Creek near Shafter.
- The Mexican Treasure of Cariza Pass: In or near Cariza Pass.

Real County

- The Lost Mine of Bull Head Mountain: Somewhere on Bull Head Mountain near Camp Wood.
- The Lost Silver Mine of the Frio River: In the vicinity of Frio River north of Leakey.
- The Lost Mine of Camp Wood: Located near the site of the old San Lorenzo Mission.

Reeves County

- Montezuma's Treasure: The legendary Montezuma Treasure has been placed in Pecos County, Reeves County, Stonewall County, and Williamson County, although the legend is hazy on specific location(s).
- La Mina Perdida: Said to be located along the Pecos River in the northeast corner of Reeves County.
- Maximilian's Lost Treasure: Ten wagon loads of gold, silver, and jewels buried at Castle Gap, west of what was known as the Horsehead Crossing of the Pecos River, north of Pecos and possibly near US 285.

Refugio County

- The Pirate Treasure of False Live Oak Point: In the vicinity of Refugio, probably at False Live Oak Point.

Sabine County

- Fowler Hamilton's Treasure: Somewhere near the Sam Rayburn Reservoir.

San Augustine County

- The Spanish Treasure of the Attoyac River: Located on east bank of Attoyac River near San Augustine.

San Patricio County

- The Lost Treasure of the Casa Blanca: In the Nueces River area above Corpus Christi in San Patricio or Nueces County.
- Enrique De Villareal's Lost Mine: In the vicinity of Edroy.
- Sterling Dobie's Treasure: Located near Mathis.
- The Blind Rancher's Treasure: Along the Nueces River south of Odem.
- The Mexican Treasure at Round Lake: At Round Lake near Mathis.

San Saba County

- Beasley's Silver Cavern: A cave thought to be located along San Saba River.

Starr County

- Pancho Villa's Treasure: Said to have been hidden somewhere near Roma.
- The Mexican Treasure of Roma: In the vicinity of toll bridge across the Rio Grande near Roma.

Stephans County

- A Sam Bass Cache: In the vicinity of Breckenridge.
- The Lost Mexican Caravan Treasure: Somewhere near Breckenridge.

Stonewall County

- Montezuma's Treasure: The legendary Montezuma Treasure has been placed in Pecos County, Reeves County, Stonewall

County, and Williamson County, although the legend is hazy on specific location(s).

- The Golden Inca Sun-God Treasure: Hidden somewhere on the Brazos River near Aspermont.
- The Lost Aztec Mission Treasure: Said to be hidden near Aspermont, and may be a variation on the stories of the preceding two sites, although the Incas were a Peruvian and South American culture, while the Aztecs, of whom Montezuma was a king, lived in neighboring Mexico.
- The Spider Rock Treasure: Said to be hidden near a place known as Spider Rock near the Double Mountain Forks of the Brazos River.

Tarrant County

- The Lost Treasure of Benbrook Lake: In the vicinity of or in Benbrook Lake near Fort Worth.
- The Riddles Ranch Treasure: Located on or near the Old Riddles Ranch near Fort Worth.

Taylor County

- The " '49ers" Treasure at Buffalo Gap: At Buffalo Gap south of Abilene.

Terrell County

- The Mexican Cave Treasure: Located near the confluence of Independence Creek and the Pecos River.

Tom Green County

- The Grandfather Clock Treasure: Hidden at or around Fort Concho near San Angelo.
- The Stagecoach Robbery Treasure in Rattlesnake Cave: Said to have been hidden in the cave ominously known as Rattlesnake Cave, near old Fort Concho west of San Angelo.

Travis County

- Treasures of Sam Bass: In the vicinity of McNeil.
- The Cave Treasure of Dagger Hollow: In a cave in Dagger Hollow on the Colorado River northwest of Austin.
- The Mexican Payroll Treasure at Shoal Creek: In the Colorado Hills near Shoal Creek.

Upshur County

- Ebenezer Bolton's Treasure: Said to have been hidden north of Gilmer.
- The Mexican Lost Gold on Little Cypress Creek: Probably on Little Cypress Creek north of Gilmer.

Upton County

- The Butterfield Stage Treasure At Castle Gap: At Castle Gap (not to be confused with the Castle Gap in Reeves County) near King Mountain.

Uvalde County

- Hoffman's Lost Lead Mine: North of Sabinal.
- The Comanche Lost Silver Mine of Frio Canyon: In Frio Canyon west of Uvalde.
- The Texas Ranger's Lost Quicksilver Mine: Located near Sibinal.

Val Verde County

- Montey Veronica Rodriguez's Treasure: In the vicinity of Shumla.
- The Bandit Treasure of Seminole Hill: Located near Comstock.
- The Lost Treasure of Pecos Canyon: In Pecos Canyon near Dorso.
- The Treasure on Mud Creek: Probably on Mud Creek near Del Rio.

Victoria County

- Moro's Treasure of Buena Vista: Somewhere along the Guadalupe River south of Victoria.

Ward County

- Chief Yellow Wolf's Treasure: Reported to have been hidden in the vicinity of Monahans.
- The Wagon Train Treasure at Willow Springs: At Willow Springs near the historical marker for the Monahans-Kermit Highway (now Route 18).

Webb County

- The Buried Treasure of Eagles' Nest: Located near Aguilares.
- The Ship Ranch Treasure: At Ship Ranch near Aguilares.

Williamson County

- Montezuma's Treasure: The legendary Montezuma Treasure has been placed in Pecos County, Reeves County, Stonewall County, and Williamson County, although the legend is hazy on specific location(s).
- The Spanish Pack Train Treasure of Leander: Located near Leander.
- The Mexican Silver Mine at Round Rock: Somewhere south of Round Rock.
- The Dead Swede's Gold of Salado Creek: Along Salado Creek in northern Williamson County.

Winkler County

- The Missouri Wagon Train Treasure: Said to have been cached in the vicinity of Willow Springs northeast of Monahans.

Wise County

- A Sam Bass Treasure: Hidden near Cave Hollow.
- H. C. Routh's Treasure: Said to have been buried on old Routh Ranch near Decatur.

Young County

- Padre and Duke's Treasure: At the site of old Fort Belknap.
- The Monterrey Raid Treasure: Loot probably taken in the raid on Monterrey, Mexico is thought to be hidden in the vicinity of old Fort Belknap.

Zavala County

- The Treasure of Turkey Creek: Said to be buried at the place where the old road from Uvalde to Las Moras crosses Turkey Creek.
- The Treasure of Loma De Sauce: Located on the Nueces River, southeast of La Pryor.
- The Treasure of Espantosa Lake: At Espantosa Lake, near Crystal City.

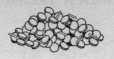

11

THE TREASURES OF THE ROCKIES

MONTANA

Before Montana renicknamed itself the "Big Sky Country" in the 1960s, it was known as the "Treasure State," and Butte, once the second largest city (after Denver) in the Mountain West, is built on what is still referred to as "The Richest Hill on Earth." The gold and silver in Butte's fabled "pit" are no longer viable commercially, but Butte is still one of the nation's top-producing copper mining areas. In the nineteenth century, and indeed well into the twentieth, Montana had the reputation of being the wildest state in the Wild West. The main street in the state capital is still known officially as Last Chance Gulch.

With the prospectors and the mining kings came the bandits and the robber barons. The most notorious bandit in Montana's history was Henry Plummer, who was to the erstwhile Treasure State what Jesse James was to Texas and the Plains. A notorious bank robber, and ironically, the former sheriff of Nevada City, California, Plummer managed to get himself appointed sheriff of Virginia City, Montana. It was an official position that was most useful to an outlaw, and he appointed his gang members as his deputies.

By the end of 1863, disgruntled citizens formed a Committee of Vigilance to counteract Plummer's lawlessness. The

Vigilantes ultimately caught up with Plummer and ended his reign of hooliganism, but his reported caches of loot are rumored to be hidden throughout the central and western parts of the state.

In addition to the lost and hidden stashes of Montana's outlaws, there are still rumors circulating about lost mines, although many of these are in virtually inaccessible corners of a state that prides itself on being largely wilderness.

Of the fifty states that we contacted for clarification of laws governing treasure hunting, Montana was one of eleven that supplied useful information. The Montana State Antiquities Act of 1978 is the only law which affords direct protection to some kinds of cultural resources on state-administered lands. The individual state agencies, however, have departmental regulations concerning access and use of the various state lands.

Montana state lands are not the same as federal lands as legal access to state lands by the public is required.

Specifically, the Montana Department of Natural Resources and Conservation requires purchase of a Recreational Use License for general recreational use of state land. Although general recreational use includes most forms of recreation, it does not include the collection of valuable rocks/minerals, collection or disturbance of archaeological, paleontological, or historical sites or artifacts, tree cutting/wood gathering, or trapping. Furthermore, the license does not grant or imply right of access.

Only those state lands which are legally accessible and have not been closed or restricted for recreational use may be utilized. Legally accessible state lands are those lands which can be accessed from adjacent public roadways, public land, and public waterways. It also includes access from adjacent private land if permission of the landowner is secured.

The two other methods by which legal access of state lands can be acquired is through a lease of the surface and/or mineral rights, or through issuance of a Land Use License. It is typically through a Land Use License that archaeological and paleontological reconnaissance and data recovery are carried out. As mandated in the Montana State Antiquities Act, however, all archaeological, historical, and paleontological sites and materials within state lands are the property of the state and will continue to be the property of the state even after removal from their original context. Further, only individuals, institutions, or agencies that meet specific academic criteria can be issued a lease or Land Use License for the sole purpose of looking for or removing archaeological, historical, and paleontological sites and materials from state lands.

As previously noted, the Montana State Antiquities Act [MCA 22-3(4)] outlines the procedures for legally, intentionally disturbing, removing, modifying, or restoring cultural (artifacts and/or archaeological/historic sites) or paleontological resources located within state lands. An artifact is deemed here as a tangible product of human behavior. An archaeological or historic site is defined here as a location identified in written documents where human activity took place and/or that contains the physical remains of past human activities. Paleontological resources refer to the remnants of fossilized plants and animals.

A heritage property is defined in MCA 22-3-421(2) as "any district, site, building, structure, or object located upon or beneath the earth or under water that is significant in American history, architecture, archaeology, or culture."

No person may excavate, remove, or restore any heritage property or paleontological remains on lands owned by the state without first obtaining an antiquities permit from the historic preservation officer. Antiquities permits are to be granted only after careful consideration of the application for a permit and after consultation with the appropriate state agency. Permits are subject to strict compliance with the following guidelines:

Antiquities permits may be granted only for work to be undertaken by reputable museums, universities, colleges, or other historical, scientific, or educational institutions, societies, or persons with a view toward dissentination of knowledge about cultural properties, provided no such permit may be granted unless the historic preservation officer is satisfied that the applicant possesses the necessary qualifications to guarantee the proper excavation of those sites and objects that may add substantially to man's knowledge about Montana and its antiquities.

All heritage properties and paleontological remains collected under an antiquities permit are the permanent property of the state and must be deposited in museums or other institutions within the state or loaned to qualified institutions outside the state, unless otherwise provided for in the antiquities permit.

No person may reproduce or falsely identify any heritage property or paleontological remains with the intent to sell the property or remains as an original. No person may sell any heritage property or paleontological remains with the knowledge that the property or remains have previously been collected or excavated in violation of MCA 22-3.

One obvious and officially acknowledged problem with applying the Antiquities Act to money or valuables cached and subsequently lost or abandoned on state lands prior to 1945 is that a case must be made to demonstrate that the buried treasure is associated with a person or event which is an important aspect of the local, regional, or national history, or that the buried treasure is considered important in addressing current research questions in history, archaeology, or paleontology. At the same time, of course, such resources are considered property of the state and cannot be removed by the public without proper authorization. Proper authorization for access of state lands and search and removal of buried treasure on state lands must ultimately come from the relevant land administering agency, such as the Department of Natural Resources and Conservation, Montana Department of Transportation, Montana Department of Fish, Wildlife, and Parks, State Department of Military Affairs, Department of Corrections, Health and Human Services, etc.

In 1991 the Montana State Antiquities Act was amended to provide protection to human burial sites and/or remains and associated artifacts. That amendment is listed as MCA 22-3(8) and provides severe penalties for disturbing grave sites and human remains.

For more information, contact

Department of Natural Resources
Division of Trust Land Management
1625 Eleventh Avenue
PO Box 201601
Helena, MT 59620-1601

Beaverhead County

- The Lost Gold of Sheep Creek Station: At the place once known as Sheep Creek Station near Beaver Head Rock.

Big Horn County

- The Lost Cabin Mine: Said to be somewhere in the drainage of the Big Horn River.
- The Treasure of the Little Big Horn: Somewhere on the Little Big Horn River, possibly near the Little Big Horn National Battlefield, which marks the site of the legendary defeat of the flamboyant antihero, Lieutenant Colonel George Armstrong Custer in June 1876.

Cascade County

- Henry Plummer's Lost Treasures: In the vicinity of Sun River.
- The Lost Mine of Jefferson Creek: Located near Jefferson Creek.
- The Buckskin Bag Treasure: Said to have been hidden in the vicinity of Dearborn.
- The Neihart Mine: Thought to be located on Carpenter Creek near Neihart.

Chouteau County

- George Keise's Lost Mine: Somewhere near Virgelle.

Deer Lodge County

- The Lost Springer Lode: Reported to be south of the town of Deer Lodge, which is in Powell County.

Fergus County

- Charlie Keyes's Lost Mine: Reported to be below the mouth of the Judith River.

Flathead County

- The Lost Mine of Moose Peak: Located somewhere on Moose Peak, north of Whitefish.

Gallatin County

- The Lost Rea Mine: Somewhere in the Gallatin Mountains in either Park or Gallatin County.
- The Red Bridge Treasure: Located near the site of the Red Bridge, near Bozeman.

Garfield County

- The Lost Indian Gold Mine: Thought to be in the Mosby area.
- The Lost Keyes Mine: In the vicinity of Piney Buttes.

Glacier County

- The Crazy Woman Lost Mine: Located near St. Mary, on the Blackfeet Indian Reservation near Glacier National Park.

Granite County

- Henry Plummer's Hidden Caches: Reported at Centennial Valley in Granite County; near Gregson in Silver Bow County and near Bonner in Missoula County, as well as near Rome Lake, Rock Hills, Haystack Butte, Ford Creek, and Ruby Valley in Madison County.
- The Robbers' Treasure of Bearmouth: In the vicinity of Bearmouth.
- The Treasure of Chinese Grade: Located near Drummond in northern Granite County.
- The Lost Gold Dust of Garnet: In the vicinity of Garnet, an area noted for the prevalence of garnets, gemstones similar to, but of lesser value than, rubies.
- Two Sleeps Lost Mine: Thought to be near Mosey, although

the name implies that it is a two- to three-day hike from somewhere, possibly from Anaconda or Philipsburg.

Hill County

- The Indian Placer of Fort Assiniboine: In the vicinity of Fort Assiniboine southwest of Havre in a stream or river with a wide enough flood plane to have once made placer mining profitable.

Lake County

- The Abandoned Cabin Lost Mine: Located near the top of Columbia Mountain in the Mission Range.
- The Lost Springer Mine: Thought to be somewhere in Lake County, although "Springer" sites are also attributed to Powell and Deer Lodge counties.

Lewis and Clark County

- The Lost Mine of Straight Creek: Located on Straight Creek in the vicinity of Camp Williams in the Lewis and Clark Mountains.
- Nepee's Lost Mine: Located near Rogers Pass on Route 200, east of Lincoln and possibly in the Lewis and Clark National Forest (*see also* Phillips County).
- The Lost Keise Mine: In the mountains near Helena.
- The Lost Mines of Helena: There are many such sites reported near the state capital and throughout the county.
- The Treasure of Ford Meadow: At Ford Meadow, about 15 miles west of Augusta.

Madison County

- Henry Plummer's Hidden Caches: Reported at Centennial Valley in Granite County; near Gregson in Silver Bow County and near Bonner in Missoula County, as well as near Rome Lake, Rock Hills, Haystack Butte, Ford Creek, and Ruby Valley in Madison County.

Mineral County

- The Chinese Cache of China Gulch: Reported to be hidden off Cayuse Creek in China Gulch near Superior.

Missoula County

- Cy Skinner's Treasure: Alleged to have been buried on an island in the Clark Fork River near, but not necessarily in, the city of Missoula, although since many of the islands are submerged in high water or in floods such as those of 1964, buried objects are likely to have been washed downstream.
- Henry Plummer's Hidden Caches: Reported at Centennial Valley in Granite County; near Gregson in Silver Bow County and near Bonner in Missoula County, as well as near Rome Lake, Rock Hills, Haystack Butte, Ford Creek, and Ruby Valley in Madison County.
- The Lost Mine of Robinson Bar: Located near Robinson Bar in the Clark Fork River in the vicinity of Bonner.
- The Gold of Fort Fizzle: In the Lolo National Forest near the location of old Fort Fizzle, a US Army installation so-named because it failed to stop the Nez Perce advance through the Bitterroot Valley in 1877.

Park County

- The Lost Rea Mine: Located somewhere in the Gallatin Mountains in either Park or Gallatin County.
- David Weaver's Lost Placer: Located on the Yellowstone River near Emigrant Peak.
- The Miner's Bootful of Gold: Rumored to be in the vicinity of Cooke City.
- The Stage Robber's Treasure: Said to be hidden near Livingston.

Phillips County

- Kid Curry's Treasure: Hidden somewhere west of Malta.
- The Lost Nepee Mine: Somewhere in Phillips County (see also Lewis and Clark County).

Powell County

- The Lost Springer Lode: Reported in Deer Lodge County, but also indicated to be south of the town of Deer Lodge, which is in Powell County.

Roosevelt County

- The Lost Gold of Poplar Creek: Reported to be somewhere near Poplar Creek, on the Fort Peck Indian Reservation.

Silver Bow County

- Henry Plummer's Hidden Caches: Reported at Centennial Valley in Granite County; near Gregson in Silver Bow County and near Bonner in Missoula County, as well as near Rome Lake, Rock Hills, Haystack Butte, Ford Creek, and Ruby Valley in Madison County.

Yellowstone County

- The Horse Thief's Treasure: Reported to be hidden in the vicinity of Billings.

IDAHO

Geographically, Idaho is two states. The panhandle in the north is characterized by deep canyons and heavily forested mountains, while the southern part of the state is mostly a vast open space. This latter area is in turn characterized by an agricultural east, a virtually uninhabited west, and a huge area of lava beds in between.

The lava beds constitute an almost impenetrable labyrinth with essentially no roads, even to this day. As such, the lava beds have traditionally provided an excellent hiding place for stolen loot. The steep canyons of the central panhandle also present a terrain that has proven useful for this purpose. Owyhee County in the southwestern corner of Idaho has the distinction of being one of the least populated and least visited

corners of the continental United States, an area the size of
Connecticut with fewer than a half dozen settlements and only
one paved road.

While tales of stolen treasure circulate throughout the state,
there are plenty of tales of lost mines throughout the western
part of the state, particularly north of Boise. The first gold
rush occurred in 1860 in the area around Orofino (which literally means ''fine gold) in and around what are now Clearwater and Idaho Counties. By 1863, there were placer mining
operations around Boise in Ada and Boise Counties. Many of
the persistent rumors of lost mines, as well as of stolen plunder, are associated with the central panhandle, and are in National Wilderness areas administered by the US Forest Service.

Ada County

- David Levy's Treasure: Said to be located in the vicinity of
  Pock Canyon.

- The Boise Stagecoach Robbery Treasure: Hidden somewhere on the south side of the Boise River near the city of
  Boise.

- The Buried Treasure of Rocky Canyon: Located in Rocky
  Canyon near Boise.

- The Treasure of Kuna Cave: Said to be located in a cave
  west of the Owyhee River near Kuna.

Bannock County

- The Gold of Portneuf Canyon: Said to be hidden in Portneuf
  Canyon near Pocatello.

- The McCammon Stage Robbery Treasure: Hidden somewhere near McCammon.

- The Treasure of Robber's Roost: Possibly related to the treasure referred to in the previous entry, this one is said to be
  located at the place known as Robber's Roost in the vicinity
  of McCammon.

Bear Lake County

- The Dying Bandit's Treasure: Said to have been hidden near Montpelier, about 12 miles north of Bear Lake.

Benewah County

- The Indian Lost Mine of Clearwater Creek: Said to be located in the vicinity of the Little North Fork of the Clearwater River.
- The Spotted Louie Mine: Located somewhere near Spotted Louis Creek.

On or near the border of Benewah and Shoshone

- The Lost Spanish Silver Mine: Said to be located in the vicinity of the St. Marie's River, southeast of St. Marie's in Benewah County or Shoshone County.
- Virgil Brumbach's Treasure: Said to be located in Soldier's Canyon, east of St. Marie's, possibly near St. Joe and possibly in the Idaho Panhandle National Forest.

Bingham County

- Blackie's Treasure: Said to be located in the vicinity of Taber.
- The Lava Fields Buried Treasure: Somewhere in the lava beds across Snake River from Blackfoot.
- The Lost Wagon Gold: Said to be located somewhere in the vast and difficult terrain of the lava beds between Arco in Butte County and Blackfoot in Bingham County.

Boise County

- George Wilson's Lost Mine and Treasure: Located near Ophir Creek and Centerville in the Boise National Forest.
- The Hoodoo Gold: Somewhere in the Boise Basin of the Boise National Forest, near Idaho City.

- The Lost Cleveland Mine: In the Boise Basin near Idaho City.
- The Stagecoach Robbery Treasure of Idaho City: Said to be located in the vicinity of Grimes Creek near Idaho City.

Bonner County

- Mieser's Treasure of Hope: Located near the town of Hope on Route 200, overlooking Lake Pend Orielle.

Bonneville County

- The Lost Trail Creek Mine: Probably near Trail Creek in the Idaho Falls region.
- The Policeman's Lost Mine: Said to be located in the vicinity of Irwin, near the Snake River on US 26, possibly in the Targhee National Forest.
- The Stagecoach Treasure of Bear Island: Hidden on Bear Island in the Snake River.

On or near the border of Bonneville and Madison Counties

- The Treasure of Kelley's Canyon: In Kelley's (or Kelly) Canyon, near Heise (aka Heise Hot Springs), which is on the north side of the Snake River at the Bonneville-Madison County line north of US 26.

Butte County

- The Highwayman's Treasure: Said to be cached on Root Hog Divide.
- The Lost Bullion on Birch Creek: Located on Birch Creek near Lone Pine.
- The Lost Gold of the Lava Beds: Stashed in the nearly impenetrable labyrinth of the Lava Beds near Big Southern Butte.
- The Treasure of the Arco Desert: Possibly a variation on the

previous entry, it is hidden in the "Arco Desert" (the Lava Beds) near Big Southern Butte.

- The Stolen Pack Train Treasure: Said to be hidden near Arco, probably in the Lava Beds, but possibly in the Challis National Forest.
- The Treasure of the Old Man of the Craters: Located near Old Man Rock in Craters of the Moon National Monument, about 20 miles southwest of Arco on US 20, US 26, and US 93.

Camas County

- The Roberts Lost Mine: Said to be located in the vicinity of Corral, near US 20.
- The Lost Sheepherder's Mine: As with the above, it is near Corral.

Canyon County

- The Saddle Tramp's Lost Ledge of Gold: Located on Squaw Creek, upstream from the Snake River near Caldwell.

Caribou County

- Caribou Lost Placer: On a stream or stream bed in the Caribou Hills near Soda Spring.
- The Treasure of Soda Springs: Located on the Bear River near Soda Springs, possibly at or across the Bear Lake County line.

Cassia County

- The City of Rocks Treasure: Near the City of Rocks National Reserve in the Albion Mountains near Almo.

Clark County

- The Buried Bandit Treasure: Said to be buried near Lidy Hot Springs.

- George Ives's Treasure: Located near Lidy Hot Springs.
- Texas Jack's Lost Mine: Located somewhere on Birch Creek.
- The Bear Hunters' Lost Mine: Still where the hunters discovered it, somewhere on Heart Mountain.
- The Lost Tenderfoot Mine: Said to be located in the vicinity of Spencer, south of the Continental Divide on Interstate 15.
- The Treasure of the Henry Plummer Gang: The legendary outlaw associated with neighboring Montana is said to have stashed some plunder in Beaver Canyon, near Spencer and about 10 miles south of the Montana line at Monida Pass.

Clearwater County

- The Lost Isaac Mine: Located on or near Coolwater Mountain, near Orofino, possibly on the Nez Perce Indian Reservation.

Custer County

- The Lost Mine of Lost River Mountain: On Lost River Mountain (or in the Lost River Range) north of Mackay.
- Isaac T. Swim's Lost Ledge: Said to be somewhere south of Bonanza and Challis, probably in the Sawtooth National Recreation Area or the Challis National Forest.
- The Lost Treasure of Bayhorse: At or near Bayhorse (or Bay Horse) near Challis.

Fremont County

- James Locket's Treasure: Located on or near Buffalo Creek.
- Ogilvie's Lost Mine and Treasure: Said to be located in the vicinity of Camas Creek (which is largely contained within Clark County).
- The Packer's Treasure: Located near Rea.
- The Robber's Grave Treasure: Said to be located in the vicinity of Rea.
- The Stage Holdup Treasure: Hidden in the Teton Basin in

southeast Fremont County, possibly in the Targhee National Forest.

- The Treasure of Sawtelle Peak: Located on Sawtelle Peak near Rea.

Idaho County

- Siawhia's Gold: South of Burgdorf and east of Warren near the Valley County line in the Payette National Forest.
- The Idaho Blue Bucket Treasure: On Dry Diggings Ridge near Warren in the Payette National Forest.
- The Lost Mine on Ruby Mountain: Somewhere on Ruby Mountain.
- The Pack Horse Lost Mine: Said to be located in the vicinity of Stites, south of Kooskia, possibly on the Nez Perce Indian Reservation.
- The Pack Train Treasure: Said to be somewhere between White Bird and Dixie in the extremely difficult terrain of the Nez Perce National Forest, where most of the few roads that exist are seasonal, unimproved roads and the steep canyons offer virtually no level ground.
- The Timber Fire Lost Mine: Somewhere in the Locksa Basin of the Clearwater National Forest.
- The White Bird Black Tail Cache: Located near White Bird, south of Grangeville on the Salmon River, and about 12 miles south of White Bird Hill, a pass crossed by US 95.
- William Rhodes's Lost Mine: Said to be south of Weippe, probably on or near Lolo Creek.

Near or on the border between Idaho County and Shoshone County

- Chief To-Moot-Sin's Indian Post Office Cache: Near the Post Office of the same name along the Lolo Trail near Cayuse Creek and US 12.

Jefferson County

- The Dead Horse Treasure: Located on Camas Creek in northern Jefferson County, possibly in the Camas National Wildlife Refuge.
- The Stagecoach Treasure on Camas Creek: Said to be located in the vicinity of Camas Creek, possibly near or across the Clark County Line.
- The Treasure of Mud Lake: At Mud Lake, north of the town of Mud Lake and west of Interstate 15.

Kootenai County

- Charles Wilson's Lost Mine: Said to be somewhere in the Coeur d'Alene Mountains, probably in the Idaho Panhandle National Forest.
- Jack Breen's Lost Diggings: Said to be in the hills near Hayden Lake and the nearby town of Hayden Lake, north of Coeur d'Alene and probably in the Idaho Panhandle National Forest.
- The Hahn Treasures: Said to be located at two widely separated sites in Idaho, with one being near Hayden Lake in Kootenai County, and the other near Gilmore Summit and Leadore in Lemhi County.
- Peg Allen's Lost Mine: Located on Latour Creek near the historic Cataldo Mission in Cataldo.
- The Lost Cabin Mine: Said to be in the Coeur d'Alene Mountains near the Cataldo Mission.
- The Lost Gold Mine of Fourth of July Canyon: In the Coeur d'Alene Mountains near Cataldo Mission, and possibly near or confused with, the previous entry.
- Richard Owens's Post Falls Diamond Mine: Located near Post Falls, which is on Interstate 90, about 5 miles from the Washington State line.
- The Post Falls-Huetter Cache: Said to be located in the vicinity of Coeur d'Alene or Post Falls, which are about 15 miles apart and linked by numerous roads, including Interstate 90.
- The Lost Mine of Taylor Peak: Said to be on Taylor Peak.

Latah County

- The Lost Mine of Paradise Creek: Located on Paradise Creek near Moscow on US 95.

Lemhi County

- The Birch Creek Stage Robbery Treasure: Probably hidden on or near Birch Creek.
- The Hahn Treasures: Said to be located at two widely separated sites in Idaho, with one being near Hayden Lake in Kootenai County, and the other near Gilmore Summit and Leadore, 50 miles south of Salmon on Route 28 in the Lemhi River Valley.
- The Stolen Stage Treasure: Said to be hidden near Leadore, in the Lemhi River Valley, or in the adjacent Lemhi and Beaverhead Mountain Ranges that parallel the valley.
- The Lost Mine on Saddle Mountain: Located on Saddle Mountain near Salmon.
- The Squawman's Lost Mine: Located on the Salmon River near Shoup in the Salmon National Forest.

Lincoln County

- The Treasure of Big Wood River: Said to be hidden in the Shoshone Ice Caves which are within a mile of the Big Wood River and near Route 75 at the northwestern corner of the lava beds.

Nez Perce County

- The Lost Tunnel of Gold: Probably located in a cave, the site is east of the Snake River near Waha.
- The Lost Wheelbarrow Mine: Somewhere in the Hoodoo Mountains, possibly on the Nez Perce Indian Reservation.
- The Robber's Treasure of Lewiston: Located near Lewiston and possibly near the Snake River.

Owyhee County

- The Owyhee Lost Placer: On a stream or stream bed in the vicinity of the Owyhee Mountains, in remote Owyhee County, which encompasses an area the size of Connecticut with fewer than a half dozen settlements and only one paved road (Route 51).

Shoshone County

- The Gold of Sly Meadows: Located at Sly Meadows near Bond Creek.
- The Lost Sheepherder Mine: Said to be in Squaw Meadows near Bond Creek.
- The Lost Sluice Boxes: Located on Placer Creek, the boxes themselves have probably deteriorated or rotted away over the years, but the placer may have value.

Teton County

- The Treasure of the Teton Basin: Located near Felt on Route 32.

Twin Falls County

- Brown's Bench Treasure: Said to be located in the vicinity of Rogerson, possibly near Salmon Creek or the Salmon Creek Reservoir.

Valley County

- The Lost Mine of Whitehawk Basin: Located on or near Whitehawk Creek.

WYOMING

While it is less well known for its gold and silver than Montana to the north or Colorado to the south, Wyoming has had its share of profitable mines. In addition, Wyoming's treasure lore contains the famous names of explorers like Jim Bridger and bandits like Butch Cassidy, who stashed their treasures in the state for a tomorrow that never came.

Cassidy, whose real name was George Leroy Parker, had joined the outlaw group known as the Hole-In-The-Wall Gang and eventually became its leader. This gang eventually joined Kid Curry's "Wild Bunch," and Cassidy became the leader of the "Bunch" as well. Both gangs were active from Montana to New Mexico, particularly in Wyoming and Colorado, whose rocky hills still hold their plunder.

Albany County

- The Downey Lost Mine: Thought to be near Centennial.
- Ezra Lay's Lost Treasure: In the vicinity of Rock River.

Carbon County

- The Old Shoe Lost Mine: In the mountains, possibly the Shirley or Medicine Bow Range, somewhere in Carbon County.
- James Shaw's Lost Mine: Located near Medicine Bow, possibly the Shirley or Medicine Bow Range.
- Outlaw Treasure: In the vicinity of Baggs in southwestern Carbon County.

Crook County

- The Canyon Springs Stage Robbery Treasure: Said to be hidden in the Black Hills, probably north of Sundance in what is now the Black Hills National Forest.

Fremont County

- Old Ike's Lost Mine: In the vicinity of Lander.
- The Oregon Trail Cache: Located on Sweetwater Creek in the Antelope Hills.
- Butch Cassidy's Treasure: Reportedly cached somewhere on the Wind Indian Reservation in Fremont County, and/or in

the Wind River Mountains which run through both Fremont and Sublette Counties.

Goshen County

- Bandit Treasure: Hidden in the vicinity of old Fort Laramie.

Johnson County

- The Lost Gold of the Seven Swedes: Thought to be located near Buffalo.

Lincoln County

- The Big Piney Treasure: In the vicinity of Smoot.

Natrona County

- The Lost Shovel Lost Mine: Thought to be near Alcova, the mine in probably easier to find than the lost shovel, but the two are probably close to one another.
- The Lost Soldier Mine: In the vicinity of Arminto.
- The Lost Cabin Mine: Somewhere in the Big Horn Mountains.

Park County

- The Cabin Creek Lost Mine: Located on Thoroughfare Butte, possibly near Cabin Creek.
- The Dutchman's Lost Mine: Reported to be on the South Fork of Yellowstone River. (As with the legendary Lost Dutchman Mine in Arizona, the term "Dutchman" probably implies a German, or "Deutsche" man.)

Platte County

- Jack Slade's Lost Treasure: Thought to be near Guernsey.
- The Buried Treasure of Sawmill Canyon: In the vicinity of Guernsey.

Sheridan County

- Jim Bridger's Lost Placer: Located on a stream or stream bed somewhere in the Big Horn Mountains.

- The Lost Mine on Bald Mountain: Somewhere on Bald Mountain.

Sublette County

- Butch Cassidy's Treasure: Reportedly cached somewhere on the Wind Indian Reservation in Fremont County, and/or in the Wind River Mountains which run through both Fremont and Sublette Counties.

- The Paymaster's Lost Treasure: Located near Big Piney.

Sweetwater County

- Big Nose George's Treasure: Hidden in the vicinity of the place known as Point of Rocks.

- The Dead Train Robber's Treasure: Located near Rock Springs.

Teton County

- The Bandit Treasure of Jackson Hole: Located in the Jackson Hole area west of Jackson and US Routes 26, 89, and 191.

- The Stage Robbery Treasure: Reportedly hidden near the town of Jackson or in the Jackson Hole area (this may be a variation on the previous entry).

Weston County

- Stagecoach Treasure: Reported to be in the vicinity of Newcastle.

COLORADO

Throughout the second quarter of the nineteenth century, while other areas of the West were being settled, or being crossed by droves of emigrants en route to Oregon and California, Colorado remained largely untouched. The Rocky Mountains which both bisect and symbolize Colorado also formed a barrier that diverted the emigrant trains to the north and south. Santa Fe was a major trading center and territorial capital for nearly 250 years before Denver was more than a tent pitched on the South Platte River.

Yet those same Rocky Mountains held a silent treasure that would transform Colorado. Gold was discovered here in 1858, ten years after the spark that launched the California Gold Rush, and a similar stampede ensued. Central City, Leadville, and Denver became overnight boom towns as first gold, then silver, poured out of the mountains. The US government even opened a mint in Denver to stamp out coins as close as possible to the source of the glittering metals.

Colorado probably had more mines—and now *has* more "lost" mines—in a concentrated area than any other state except California. In the southwestern part of the state there are major concentrations in the San Juan National Forest area that stretches across Montezuma, La Plata, San Juan, Archuleta, and adjacent counties.

West of Denver, Larimer and Summit Counties are particularly rich, and the southeastern counties of Costilla, Huerfano, and Las Animas are home to many legends of lost loot. These include the mysterious Arapaho Princess Treasure in the Spanish Peaks area, the Green Lost Mine near Red Wing, the myriad of lost mines on La Veta Creek near Walsenburg and the most tantalizing sunken treasure at the bottom of Bottomless Lake in Mustang Canyon.

Archuleta County

• Jim Stewart's Lost Placer: Southwest of Pagosa Springs.

• The Lost Silver Lake of Edward "Nut" Ennis: Somewhere in the San Juan Mountains.

- Kit Carson's Lost Gold Placer: Located near Little Blanco Creek, southwest of Pagosa Springs.
- The Lost Gold of Ute Mountain: Located on Ute Mountain.
- The Lost Treasure of Treasure Mountain: Located on Treasure Mountain, east of Pagosa Springs.
- The Lost Ute Mine: Located on Ute Creek, northwest of Pagosa Springs.
- The Treasure of Chimney Rock: In the vicinity of Chimney Rock.

Bent County

- The Lost Gold of Bent's Fort: At the site of Bent's Old Fort (now Bent's Old Fort National Historic Site) on Crow Creek near Las Animas.
- The Treasure of Las Animas: Located near Las Animas.

Chaffee County

- The Gold Nuggets of Lost Canyon: In the vicinity of Granite.
- The Spanish Princess Treasure: Reported to be hidden on Mount Princeton, north of Alpine.

Clear Creek County

- The Lost Treasure of Georgetown: Hidden in the mountains near Georgetown.
- The Lost Saxon Mine: In the vicinity of Berthoud Pass near Empire.
- The Gabe Espinoza Treasure: Located on or near Mount Evans.
- Douglas McLain's Lost Mine: On Hick's Mountain, west of Brookvale.

Conejos County

- Josh Thomas's Treasure: Located on Conejos Creek.
- The Lost Mine of Hidden Valley: Located near La Jara Creek.

Costilla County

- Manuel Torres's Lost Mine: Located on or near Culebra Peak southeast of San Francisco.

- White's Lost Cement Mine: Located on or near Culebra Peak southeast of San Francisco.

- The Spanish Treasure of Blanca Peak: Located on Blanca Peak north of Fort Garland.

- Juan Carlos's Lost Gold: In the vicinity of Blanca Peak (this story may be a variation on the story in the preceding entry).

- The Paymaster's Treasure of Fort Garland: Located on Trinchera Creek south of Fort Garland.

Custer County

- George Skinner's Lost Mine: On Horn Peak Mountain west of Westcliffe.

- The Cave of Gold: A cave located somewhere on Marble Mountain.

- The Treasure of Deadman Caves: The caves are said to be located near Deadman's Camp, on Deadman's Creek.

Delta County

- The Lost Pin Gold Mine: In the vicinity of Delta.

Dolores County

- The Stolen Bullion on Indian Ridge: Located on Indian Ridge in the La Plata Mountains near Rico.

- The Lost Gold of Lizard Head Pass: The pass is located on Bear Creek near Rico.

- The Treasure of Cherry Creek: Located on Cherry Creek near Rico.

Douglas County

- Albright's Lost Silver: Said to be hidden about 20 miles south of Denver in the Kinney Park area.
- The Lost Gold of Stevens Gulch: Reported to be near or in Stevens Gulch east of Strontia Springs.
- The Treasure of Devil's Head: Located near Devil's Head Rock east of Deckers.

Eagle County

- The Crazy Woman Lost Mine: In the vicinity of Mintura.
- The Lost Ore of Slate Mountain: Located on or near Slate Mountain, southeast of Eagle.

Fremont County

- Johnson's Lost Tungsten Mine: Located near Canon City.
- Royal Gorge Gold: Located in Copper Gulch near Royal Gorge.
- The Gold Tom Park Treasure: In the vicinity of Colapaxi in western Fremont County.

Garfield County

- The Bandit Treasure of Grand Valley: Located in or near Grand Valley.
- The Lost Mine of Uintah Basin: Located near Douglas Pass in northwestern Garfield County.
- Train Robbery Treasure: Hidden in the vicinity of Grand Valley.

Grand County

- The Lost Bonanza on Soda Creek: Located on Soda Creek northwest of Grand Lake in the Arapaho National Forest.

Gunnison County

- Adam Murdie's Lost Gold Mine: Located near Taylor Park and Gunnison.
- Joseph John's Lost Mine: Located near Muddy Creek south of Placita.
- The Jesse James Treasure: In Half Moon Gulch southwest of Leadville.
- The Lost Copper Mine on Meadow Mountain: Located in Crystal River Valley near Meadow Mountain.
- The Lost Mine of Snowblind Gulch: In Snowblind Gulch at the western foot of Monarch Pass.
- The Lost Nuggets on Crow Creek: Located on Crow Creek near Crested Butte.
- The Treasure of the Cement Creek Caves: The Cement Creek Caves are located near Crested Butte.

Hinsdale County

- The Crazy Swede's Lost Mine: Said to be located between Ouray and Lake City.

Huerfano County

- Alex Cobsky's Lost Mine: Located at or around La Veta Pass near Silver Mountain, north of La Veta.
- Henry Sefton's Treasure: Said to be on the Gomez Ranch in the Sangre de Cristo Mountains.
- The Huajatolla Gold: Located on Spanish Peak in the Sangre de Cristo Mountains.
- Jack Simpson's Lost Mine: At La Veta Pass near Silver Mountain, north of La Veta.
- Lost Mine of Wet Mountain Valley: In the Wet Mountain Valley of southern Huerfano County.
- Spanish Peaks Lost Mine: Located in the Spanish Peaks area of southern Huerfano County.
- The Arapaho Princess Treasure: Located near Spanish Peak between Walsenburg and Trinidad.

- The Breckinridge Huntsman Lost Mine: Somewhere in southern Huerfano County.
- The Green Lost Mine: In the vicinity of Red Wing.
- The Jasper Lost Mine: Located near Red Wing.
- The Lost Mine of Veta Creek: Located on La Veta Creek, near Walsenburg.
- The Sunken Treasure of Mustang Canyon: Somewhere at the bottom of Bottomless Lake in Mustang Canyon, near Walsenburg.
- The Spanish Mine of Apache Gulch: In Apache Gulch near Walsenburg.
- The Treasure of Spanish Fort: In the vicinity of the old Taos Trail, about 25 miles west of Walsenburg.

Jackson County

- The Rabbit Ears Lost Mine and Treasure: Located on Walden Creek near Rabbit Ears Peak.

Jefferson County

- Ben Leeper's Treasure: Located in the mountains near Golden.
- The Lost Gold on Ralston Creek Road: Located on Ralston Creek Road, between the former mining capital of Central City and Denver, the present state capital.
- The Treasure of Standley Lake: Located at or in Standley Lake.

Lake County

- Pat Kelly's Lost Mine: In the vicinity of Lake City.
- The Baby Doe Treasure: At Matchless Mine, near Leadville.
- The Frenchman's Convoy Treasure: Located near Leadville.
- The Lost Gold Mine of Lost Gulch: At Lost Gulch near Leadville.

La Plata County

- The Loma Gold: Located in the Animas River Valley near the tracks of the Durango & Silverton Narrow Gauge Railroad.
- Lone Wolf's Lost Mine: Somewhere on the western slope of Parrott Mountain.
- Milt Hollingsworth's Lost Venison Lode: In or near Root Gulch.
- Newton's Lost Gold Mine: In the San Juan Mountains near Durango.
- The Animas Treasure: In the vicinity of Bondao in south central La Plata County.
- The Buttinski Lost Gold Strike: Located near Durango.
- The Lost Josephine Mine: In the La Plata Mountains.
- The Lost Mine of Bear Creek: Located on Bear Creek, southwest of Creede.
- Thomas Estes's Lost Mine: On the southern slope of West Needle Mountain.

Larimer County

- Albert Bierstadt's Lost Lode: In the vicinity of Wild River, southwest of Estes Park, probably in Rocky Mountain National Park.
- The Hunter's Gold: Located near Wind River, southwest of Estes Park, probably in Rocky Mountain National Park.
- Jacques Borgeans's Lost Mine and Treasure: Located near Fort Collins.
- The Buried Treasure of Natural Fort: In the vicinity of Fort Collins.
- The Dying Prospector's Lost Mine: Somewhere on Specimen Mountain, northwest of Estes Park, probably in Rocky Mountain National Park.
- The Gunshop Lost Lead Mine: In the hills near Fort Collins, probably in the Roosevelt National Forest.
- The Lost Dutchman Mine of Colorado: On Black Mountain, northwest of Fort Collins. (As with the legendary Lost

Dutchman Mine in Arizona, the term "Dutchman" probably implies a German, or "Deutsche" man.)

- The Lost Dutch Oven Gold of Grand Lake: East of Grand Lake in Rocky Mountain National Park.
- The Musgrove Corral Treasure: Located near Fort Collins.
- The Treasure of Robbers Roost: At Robbers Roost, west of Virginia Dale.

Las Animas County

- Ashton B. Teeples' Lost Golden Lake: Somewhere in the Culebar Mountains near Trinidad.
- Jim Cockrell's Lost Placer: In a stream or stream bed in the vicinity of Trinidad.
- Pierre's Lost Mine: Somewhere in Las Animas County.
- The Handyman's Hidden Mine: In the Mesa de Maya area near Branson.
- The Lost Mine of Las Animas Valley: In the Las Animas Valley.
- The Lost Mine of Marble Mountain: On Marble Mountain in the Spanish Peaks region.
- The Spanish Peaks Lost Mine: Located in the Spanish Peaks area of western Las Animas County.
- The Spanish Treasure of Purgatoire Canyon: Located in Purgatoire Canyon between Las Animas and Walsenburg.
- The Treasure of Shell Canyon: Hidden in Shell Canyon, about 18 miles north of Kim.
- Whatoyah Lost Mine: In southern Animas County.

Lincoln County

- The Bandit Treasure of Clifford: Hidden somewhere east of Clifford.

Mesa County

- The Landlady's Lost Silver: Located about 10 miles northeast of Fruita.
- The Lost Pin Gold Mine: In the Grand Junction region.

Mineral County

- Curt Gardner's Lost Golden Cave: Located near Biedell, north of Creede.
- The Lost Mine of Hick's Mountain: Somewhere on or near Hick's Mountain.
- The Lost Mine of Mogate Peak: Located on Mogate Peak, near Creede.
- The Lost Mine of Treasure Mountain: On Treasure Mountain near Creede.
- The Lost Mine on Embargo Creek: Located on Embargo Creek, near Creede.
- The Three Skeletons Lost Mine: In the vicinity of Bear Creek.
- William Perkins's Lost Mine: Located near and possibly on, Sierra Blanca Peak.

Moffat County

- The Snake River Gold: Located at headwaters of Little Snake River, and, as the legend goes, on the 41st parallel.
- The Hanson Brothers' Lost Gold: In the vicinity of Maybell.
- The Lost Cemented Placer Gold: Located near Craig.
- The Lost Mine on the Yampa River: In the vicinity of Elk Springs, on the Yampa River.
- The Lost Phantom Mine: North of Dinosaur National Monument in northwestern Moffat County.
- The Treasure of Pat's Hole: At Pat's Hole in Dinosaur National Monument, Moffat.

Montezuma County

- Butch Cassidy's Treasure: Located near Powder Springs in northern Montezuma County.
- The Butterfly Lost Mine: Located on Hesperus Peak, northeast of Mancos in the San Juan National Forest.
- The Lost Joseph Mine: In the La Plata Mountains, northeast of Cortez.

- The Montezuma Treasure: West of Durango and possibly near US 160.

Montrose County

- The Treasure of the Denver Mint: Gold and silver coins cached somewhere between Crawford and Montrose, deep in a chasm formed by the Gunnison River.

Otero County

- The Wagon Train Treasure: Hidden in the vicinity of La Junta.

Ouray County

- The Lost Mine on Oak Creek: Located on Oak Creek, west of Ouray in the Uncompahgre National Forest.

Park County

- The Groundhog Lost Mine: Near Hoosier Pass, north of Fairplay in the Pike National Forest.
- The Lost Mine of Dead Man's Gulch: In Dead Man's Gulch on Kenosha Pass.
- The Reynolds Gang Treasure: Hidden in or near Hand Cart Gulch.
- The Treasure of Spanish Caves: In the Spanish Caves in the vicinity of Alma.

El Paso County

- Outlaw Treasure: Located near Manitou Springs (suburban Colorado Springs), possibly in the Pike National Forest.

Pitkin County

- The Clubfoot Lost Mine: At Root Gulch, north of Parrott Mountain.

- The Lost Mine of Meadow Creek: At McClure's Pass, west of Redstone in the White River National Forest.
- The Lost Mine of Conundrum Gulch: South of Aspen in the White River National Forest.
- The Lost Mine of Hunter Creek: Located near Aspen, on or near Hunter Creek, in the vicinity of Thimble Butte and Smuggler's Mountain.

Pueblo County

- Mike Clay's Lost Nuggets: As the story goes, they were lost somewhere in the vicinity of Pueblo.
- The Aztec's Treasures: Located near the St. Charles River.
- The Lost Mine of the Greenhorn Mountains: Located somewhere in the Greenhorn Mountains west of Pueblo.

Rio Grande County

- Phillips' Lost Lode: Said to be located west of Summitville.

Routt County

- Jim Baker's Lost Mine: Located near Clark, between Steamboat Springs and Hahn's Peak in the Routt National Forest.
- The Lost Mine on Elk Creek: Located on Elk Creek, about 10 miles east of Hahn's Peak, near the Jackson County Line in the Routt National Forest.
- The Lost Phantom Mine: Said to be in the vicinity of Steamboat Springs, probably in what is now the Routt National Forest.

Saguache County

- The Buried Treasure on Round Hill: Located on Round Hill near Pancha Pass.
- The Lost Mine in the Sangre de Cristo Mountains: Located near Cottonwood Creek in the Sangre de Cristo (Blood of Christ) Mountains.

San Juan County

- Levi Carson's Lost Mine: Located somewhere on the northern slope of West Needle Mountains.
- The Baker Brothers' Lost Ledge: Located on Coal Creek, near Silverton.
- The Treasure of Timber Hill: Located on Timber Hill, near Silverton.

San Miguel County

- Henry Sommer's Treasure: Located on Falls Creek, near Monument Mountain.
- The Lost Trail Mine: Located on Mount Wilson in the San Juan Mountains.

Sedgewick County

- The Buried Treasure of the Italian's Cave: In or near the Italian's Cave near Julesburg on the Nebraska state line.
- Jules Beni's Treasure: Reported to be in the Indian Caves southeast of Julesburg, this site is probably the same as the previous entry.

Summit County

- George Franz's Lost Mine: Somewhere in the Gore Range in the White River National Forest.
- Hill's Lost Mine: In the vicinity of Heeny.
- Horace Pullen's Lost Mine: Located near Dillon on US 6 and Interstate 70.
- John La Foe's Lost Mine: Located in the Gore Range in the White River National Forest.
- Lem Pollard's Lost Mine: Also located in the Gore Range in the White River National Forest.
- The Black Princess Mine Shaft Treasure: Located in the vicinity of Breckenridge.

- The Lost Tenderfoot Mine: Located south of Breckenridge in the White River National Forest.
- The Frying Pan Lost Mine: Located in the Gore Range, in the White River National Forest northwest of Dillon.
- The Lost Gold of Gore Range: Gore Range in the White River National Forest near Piney Lake. (This is possibly a variation on one of the previous Gore Range entries.)
- The Lost Mine of Bear Mountain: Located on Bear Mountain, near Breckenridge.
- The Lost Silver Mine of Gore Canyon: Located in Gore Canyon in the Jacques Mountains.
- The Whittler's Lost Mine: Located near Breckenridge, possibly in the White River National Forest.

12

THE TREASURES OF THE BASIN STATES

NEVADA

Today, there is probably no state that is more closely identified with the myth of striking it rich overnight than Nevada. People do, but many more go home having been struck down, struck dumb by loss, or having simply struck out. Gambling— or "gaming" as they like to say in Nevada—is by far the major industry, but it is a relatively recent industry. Before the boom in big-time casinos that began in the early 1950s, Nevada had existed only as a vast empty place that one had to cross in order to get to California. Before gaming transformed two tiny corners of the state, the only major boom— and it *was* a big one—had been the great silver rush of the 1860s that pulled millions of pounds of silver out of the great Comstock Lode in the Reno–Virginia City area and gave the Silver State its enduring—and endearing—nickname.

In terms of its physical environment, Nevada is a state of stunning, and often shocking, contrasts. Las Vegas and the Reno–Lake Tahoe area are major concentrations of people, activity, and wealth, yet in 99 percent of the state, one can look in all directions and neither see nor hear another human being. There are counties in Nevada that are larger than Portugal but contain fewer people than a New York subway train.

Nevada is the emptiest state in the continental United States, and US 50, which crosses Nevada at its mid-section is billed as the "World's Loneliest Highway" because one can drive for hours without seeing a place of human habitation. Much of this region is also extremely dangerous, a hot and waterless desert filled with poisonous snakes, spiders, scorpions, and gila monsters in summer, and drifting snow in winter.

For example, there are several alleged locations of hidden treasure in Elko and Humboldt counties where one would have to drive for the better part of a day on unimproved or unpaved roads just to get from a main highway to the town near which the treasure is said to be located. From there, it may be another day or two on foot or horseback before one can actually *start* to search for the treasure.

The Silver State is also blessed—or cursed—with the largest proportion of federal land in the West. There are National Forests and National Wildlife refuges, but there is also a great deal of Bureau of Land Management open space, and restricted military areas. The latter are entirely off limits to would-be treasure hunters. There is a US Air Force Range—mostly in Nye County—associated with Nellis AFB that is larger than Belgium, but which officially doesn't exist. The Department of Energy runs the "Nevada Test Site"—also in Nye County—with access only to those testing nuclear weapons.

Excepting the troves within the gaudy casinos and gas station slot machines, the treasure lore of the Silver State includes both the remnants of the great Comstock Lode mines, and also the often-pathetic diggings of desperate prospectors who were eternally, and ultimately fatally, just one shovelful from immense wealth.

Churchill County

- Henry Knight's Lost Gold-Lined Cave: In the vicinity of Sand Springs and Painted Hills, southeast of Fallon.
- The Chicken Craw Lost Mine: Located near Fallon.
- The Langford Treasure: Located on Chalk Mountain.
- The Shoshone Lost Mine: In the vicinity of Sand Springs east of Fallon.

Clark County

- The Black Sand Lost Mine: Located near Searchlight, 30 miles south of Las Vegas on US 95.

- Bugs Maraval's Dead Mountain Gold Cave: In the Dead Mountains, south of Searchlight and north of Needles, California.

- Lawrence's Lost Diamond Ledge: Located on the Colorado River, in the Lake Mead National Recreation Area south of Las Vegas.

- The Mashbird Lost Mine: Located in the McCullough Mountains, south of Las Vegas and between US 95 and Interstate 15.

- The Lost Devil's Peak Mine: On Devil's Peak.

- The Lost Mormon Silver Ledge: In the McCullough Mountains, south of Las Vegas.

- The Lost Platinum Ore Body: In the vicinity of Goodsprings on Route 161.

- The Lost Skillet Mine: Said to be located near Searchlight.

- The Quejo Treasure: In the vicinity of Boulder City on US 93.

- The Spirit-Cursed Gold Mine: A site not to be considered for even a moment by the superstitious, this mine is said to be located in the Dead Mountains in southern Clark County.

- The Treasure of Mountain Springs: Near Mountain Springs in the Potosi Mountains southwest of Las Vegas on Route 160.

Douglas County

- The Cody Lost Mine: In the Pine Nut Range east of US 395.

- The Empire Stagecoach Holdup Treasure: Said to be hidden near the Nevada State Prison at Carson City.

- The Stagecoach Treasure of Genoa: In the vicinity of Genoa on Jacks Valley Road between the Toiyabe National Forest and US 395.

Elko County

- John Esterly's Lost Gold Ledge: Located on or near Monument Peak.
- Swede Pete's Lost Gold Ledge: Said to be in the Ruby Mountain foothills near Elko, probably in or near the Humboldt National Forest.
- The Jarbridge Stage Treasure: Hidden north of the ghost town of Jarbridge (reachable only by way of a 100-mile, mostly-unpaved road from Rogerson, Idaho, which is on US 93).
- The Sheepherder's Lost Claim: In the vicinity of the ghost town of Jarbridge (*see* previous entry).
- The Lost Copper Lode of Mountain City: In the vicinity of Mountain City in the Independence Mountains north of Elko.
- The Lost Gold Cache on the Bruneau River: Located near Pen Rod Bridge on the Bruneau River north of the ghost town of Charleston, which is more than 20 miles on an unpaved road from Route 225.
- The Lost Silver in the Ruby Mountains: Several such sites are located in the Ruby Mountains which are mostly in the Humboldt National Forest and reachable only on foot or on unpaved roads for part of the way.
- The Missing Gold at Tuscarora: In the vicinity of Tuscarora, west of the Independence Mountains and seven miles from Route 226 on an unpaved road.
- The Pick and Shovel Lost Mine: Located near Kelly Creek.
- The Stagecoach Treasure of Williams Station: Located on Harrison Creek near Hill Beacher Road.

Esmeralda County

- Charles Lampson's Lost Gold Float: Located near Crow Springs on the old Sodaville-Tonopah Road.
- The Columbus Stagecoach Treasure: Hidden in the vicinity of Columbus.
- The Stovepipe Lost Mine: Located near Weepah.

- The Lost Gold of the Monte Cristo Mountains: In the Monte Cristo Mountains west of Tonopah and north of US 95 and US 6.

Humboldt County

- Joseph Taylor's Lost Indian Placer: In the vicinity of Ebbings.

- Old Man Barry's Lost Mine: Said to be somewhere on Pahute Peak.

- The '49er Treasure: Hidden somewhere on Disaster Peak.

- The Lost Frenchman Mine: In the Black Rock Range, near the Black Rock Desert, north of the ghost town of Sulphur, which is 60 miles west of Winnemucca on an unpaved road.

- The Lost Ledge of the Black Rock Desert: In the Black Rock Desert, between Black Rock and Mud Meadows, north of the ghost town of Sulphur, which is 60 miles west of Winnemucca on an unpaved road.

- The Lost Mine of the Little Brown Men: In the Division Peak area, northwest of Sulphur.

- The Lost Gold of the Three Lakes: Near the "Three Lakes" on Adam Peak, north of Golconda on Interstate 80.

- The Lost Mine on Buckskin Peak: On Buckskin Peak.

- The Lost Silver of Hardin City: Located near old mining town of Hardin City.

- The Lost Tenderfoot Mine: Somewhere in Humboldt County.

- The Padre's Lost Ledge: Between Buffalo Springs and High Rock Canyon.

- William Daniely's Lost Bonanza: At the site of the old National Mine.

Lander County

- The Lost Mine of The Peaks: In the vicinity of "The Peaks."

Lincoln County

- A Horse Thief's Treasure: In the Pahranagat Valley, south of Hiko near the junction of US 93 and Route 375.
- Ellen's Lost Mine: In the Highland Mountains near Pioche.
- The Lost Doublecross Mine: Somewhere in southwestern Lincoln County, probably on the restricted US Air Force bombing range, and therefore off limits. This site is possibly in or near the super-secret "Area 51" that is of much interest to UFO conspiracy theorists.
- The Mormon Caravan Treasure: Between Cave Valley and Ash Meadows, near Carp, 20 miles south of Route 317 on an unpaved road.
- The Paiute Hidden Gold: A Native American cache located near Delamar.
- The Double Cross Lost Mine: Located on Quartz Peak.

Mineral County

- The Candelaria Mine Lost Treasure: Located near Columbus.
- Jim Nelson's Lost Mines: In the vicinity of Mina, which is on US 95.
- The Lost Mine on Cedar Mountain: On Cedar Mountain, near Mina.
- Tim Cody's Lost Ledge: In the vicinity of Cedar Mountain.
- Pedro's Lost Silver: In the Excelsior Mountains, near old mining town of Candelaria in southern Mineral County.
- The Lost Quicksilver of Belleville: Located near Belleville. (Please remember that quicksilver is mercury, and mercury can be poisonous to the touch.)
- W. A. Hawthorne's Lost Mine: Located near Luning on US 95.

Nye County

- Duckett's Lost Mine: On Black Mountain in the Monitor Range, east of the ghost town of Belmont in the Toiyabe National Forest.

- Judge Whitaker's Lost Gold: In the Toquima Range west of Belmont.
- The Lost Cabin Mine: In the vicinity of Tonopah on US 6 and US 95.
- The Lost Gold Ledge of Wheelbarrow Peak: Located near Wheelbarrow Peak in the Kawich Valley which runs into the restricted US Air Force bombing range area.
- The Lost Mine of Johnnie: Located on Crescent Wash, near Johnnie, southern Nye County.
- The Lost Stovepipe Mine: Said to be in southern Nye County, it may be confused with the Stovepipe Mine in Esmeralda County.
- The Treasure of Mount Helen: On Mount Helen.

Ormsby County

- The Bank Robber's Treasure at Six Mile Canyon: In the vicinity of Six Mile Canyon, near the road from Carson City to the ghost town of Ramsey.
- The Lost Mercury in the Carson River: Said to be in the Carson River near Empire.
- The Lost Payroll of Lake Tahoe: Hidden near Lake Tahoe in the vicinity of Glenbrook.

Pershing County

- Prengle's Lost Mine: Located near Antelope Springs.

Storey County

- The Arch Treasure: In the vicinity of Virginia City.
- The Grosch Brothers' Lost Silver Mine: Located near Virginia City.

Washoe County

- The Forman Lost Lead Mine: Somewhere in the Granite Range.

- Jasper Price's Lost Gold Mine: On Slide Mountain near Reno.
- The Chinese Treasure at Pyramid Lake: Located at or near Pyramid Lake, probably on the Pyramid Lake Indian Reservation.
- The Lost Cave of Gold: Said to be somewhere in southern Washoe County.
- The Lost Golden Eagle Mine: Said to be somewhere in southern Washoe County.
- The Turkey Treasure: In the vicinity of Reno.

White Pine County

- The Pogue Station Treasure: Located near the old Pogue Station southeast of Eureka, possibly near Pinto Summit on US 50.
- The Stolen Quicksilver of Eberhardt: Located in the White Pine Mountains, near old mining town of Hamilton in southwestern White Pine County.

UTAH

Geographically, Utah is the meeting place of the defining features of the adjoining states—the Great Basin deserts of Nevada, the Rocky Mountains of Colorado, and the painted canyonlands of Arizona. The most distinctive feature of course, is the Great Salt Lake, the second saltiest body of water in the world (after the Dead Sea) and the largest land-locked body of salt water in the Western Hemisphere. Historically, Utah was lightly populated by the Ute and Shoshone people and so it remained until 1847, when the Mormons (members of the Church of Jesus Christ of Latter Day Saints) under Brigham Young selected the area east of the Great Salt Lake as a place to establish their religious community. Persecuted in the East, the Mormons chose to settle in a place as far from mainstream civilization as possible. They found it, and for a while, they lived in it. Between the time of the California Gold Rush of 1849 and the completion of the first transcontinental railroad line in 1869, the wishes of the Mormons to be a people apart were gradually diluted by the influence of non-Mormons.

The treasure lore of Utah parallels that of neighboring states. There were gold and silver mines, although they never achieved the levels of production found in the grand strikes in Montana and Colorado. Coal was, and continues to be, one of Utah's most important minerals.

The Mormons were generally law-abiding (except for events such as the Mountain Meadows Massacre, and their ignoring of the legal ban on polygamy), but with the mines and the influx of population—particularly after the arrival of the railroad—came the lawlessness associated with the surrounding states. George Leroy Parker, who would achieve notoriety under the name Butch Cassidy, got his start in Utah, and he roamed the state with his Hole-in-the-Wall Gang, and later with the Wild Bunch.

Also in the Utah lore are the lost treasures of the people who simply passed through the state. Many California-bound emigrants suffered mightily from heat and lack of water as they crossed the Great Salt Desert west of the Great Salt Lake. In 1846, the Donner Party was no exception, but they also achieved dubious notoriety later in the year as they tried to cross the Sierra Nevada in the winter. Caught in a heavy snow, they were forced to resort to cannibalism to survive. However, during their crossing of Utah, they are said to have stashed some of their valuables near Wendover for a later retrieval that would never come.

Beaver County

- The Rattlesnake Lost Mine: In the vicinity of Milford.
- The Treasure of the Indian Caves: West of Antelope Springs, about 18 miles north of Milford on Route 257 near the Millard County line.

Box Elder County

- The Corinne Stage Treasure: Hidden in the vicinity of Corinne, west of Brigham City.

- The Lost Bidwell Treasure: Said to be somewhere in western Box Elder County, northwest of the Great Salt Lake and north of the Great Salt Lake Desert.
- The Train Robber's Treasure at Bear River: Located near Bear River City, about 4 miles north of Corinne on Route 13.

Cache County

- Bandit Treasure: Located near Logan in Cache Valley.

Carbon County

- Brigham Young's Lost Gold Mine: Said to be somewhere in the Uintah Mountains, possibly near Price.

Daggett County

- The Wild Bunch Treasure: Said to be hidden in the vicinity of Brown's Hole.
- Thomas Ewing's Lost Mine : Located near Brown's Hole.

Duchesne County

- Caleb Rhoades's Lost Mine: Said to be in the vicinity of Moon Lake in the Uintah Mountains.

Emery County

- A Butch Cassidy Gang Treasure: Said to have been cached near Buckhorn Wash.
- The Army Payroll Treasure: The hidden loot is in the vicinity of Castle Dale.
- The Bank Robbery at Emery: The stolen money was hidden at Hondo Arch near Emery.

Garfield County

- The Treasure of Davis Canyon: Thought to be hidden along Paria Creek in Davis Canyon.
- The Wolverton Lost Mine: Somewhere in the Henry Mountains.

Grand County

- Jack Wright's Lost Mine: Located near Moab on US 191.
- The Golden Jesus Cave Treasure: Somewhere in the La Sal Mountains.
- The Japanese Cook's Treasure: Hidden in the vicinity of Cisco near the Colorado River south of Interstate 70.
- The Lost Caleb Rhodes Mine: Somewhere in Grand County.
- The Lost Josephine Mine: In the La Sal Mountains.
- The Train Robber's Treasure: In the Grand Valley.

Iron County

- The Lost Ledge of Gold: Said to be located near Summit on Interstate 15.
- The Mountain Meadow Massacre Treasure: About 30 miles south of Cedar City at the site of the September 1857 massacre of 137 emigrants by Mormon settlers.

Juab County

- The Lost Mine in the Sevier Desert: In the Sevier Desert near Mammoth on US 6.
- The Mammoth Miner Lost Mine: In the vicinity of Mammoth on US 6.

Kane County

- Montezuma's Treasure of White Mountain: On White Mountain near Kanab and the Arizona state line.

- The Rifle Sight Lost Mine: Said to be in the vicinity of Alton.

Millard County

- Margum Pass Lost Mine: Possibly at or near Margum Pass.
- Sheepherder Pedro's Lost Mine: Located near Morgan Pass, west of Delta.
- The Lost Mine on Cricket Mountain: Located on Cricket Mountain near Pumice.

San Juan County

- Abel Herring's Lost Gold: West of Bluff, possibly in the Valley of the Gods off Route 261.
- John Douglas's Lost Mine: In the vicinity of Mexican Hat in southern San Juan County, possibly in or near Chinle Wash, and probably on the Navajo Indian Reservation.
- Samuel Brooks's Lost Mine: In the Comb Range.
- The Lost Gold of the San Juan River: Located on the San Juan River near Mexican Hat.
- The Lost Pot Hole Placer: On Abajo Peak, north of Blanding on or near US 191.
- The Lost Uranium Strike: West of Monticello in the Abajo Mountains and/or the Manti-La Sal National Forest.
- The Spanish Treasure: Said to be hidden on Moonlight Creek.
- The Ute Lost Placer: Located in a stream or stream bed near Jim Black Basin.

Summit County

- The Temple Lost Mine: Somewhere on Gilbert Peak.
- The Spanish Lost Mine: In the vicinity of Henefer.
- Truelove Manheart's Lost Ledge: Somewhere in the vicinity of Hayden Fork.

Tooele County

- Brigham Young's Lost Gold Mine: Said to be somewhere in the Oquirrh Mountains, possibly near Tooele.

- The Crossland Lost Mine: Located near Tooele.

- The Donner Party Treasure: Said to be hidden near Wendover by the ill-fated group during their 1846 crossing of the Great Salt Desert.

Uintah County

- Blackjack Ketchum's Treasure: In the vicinity of Vernal.

- The Bullet Lost Mine: Located near Ouray, possibly in the Desolation Canyon of the Green River or on the Hill Creek Extension of the Uintah and Ouray Indian Reservation.

- The Indian Cave Treasure: In the vicinity of Kansas.

- The Lost Ewing Mine: Somewhere in the Uintah County region.

- The Mormon Lost Mine: In the Uintah Mountains north of Vernal.

- The Spanish Treasure of the Uintah Mountains: In the Uintah Mountains and/or the Ashley National Forest, which spread north and west from Uintah County into Daggett and Duchesne Counties.

Utah County

- The Nephite Cave Treasure: Located near the town of Spanish Fork on US 6.

Wasatch County

- The Dying Miner's Lost Mine: Said to be located in the vicinity of Hailstone.

- The Forest Fire Lost Lode: Located near Park City.

Washington County

- The Dixie Lost Mine: In the vicinity of Leeds, probably in the Pine Valley Mountains in what is now the Dixie National Forest.
- The Lost Gold of Hurricane Cliffs: Located at Hurricane Cliffs near the Arizona-Utah border.
- The Lost Lead of Santa Clara: Between Gunlock and Santa Clara, near Shivwits and possibly near Snow Canyon State Park.
- Utah's Ghost Gold: Located near Meadow Valley Creek.

Wayne County

- The Castlegate Payroll Treasure: Hidden in the vicinity of Hanksville.

13

THE TREASURES OF THE SOUTHWEST

ARIZONA

Arizona and New Mexico can be a wonderland of light and shadow, where the colors of the mesas drift from fiery red to deep purple in a matter of moments. The Southwest is also a place where a person can die of thirst, and his bones bleached by the sun before he is found.

Yet, out in that wonderland, there are lost riches waiting to be found. If you cock your ear, you can almost hear the wind telling you where it was hidden by this wily Apache, or that paranoid Frenchman. It's almost worth risking a venture into a hellish place where a person can die of thirst, and disappear without a trace.

As late as the second decade of the twentieth century, Arizona and New Mexico were still so remote that they had not yet achieved statehood. They were the last states of the contiguous forty-eight admitted to the Union. Even today, aside from the sprawling metropolis that surrounds Phoenix, and the hubs of activity in Tucson, Albuquerque, and Santa Fe, the Southwest is still a wild and remote land, where one can walk for hours, if not days, without seeing another human being.

The Spanish Conquistadors arrived here from Mexico in the sixteenth century, looking for gold and silver, and seeking the fabled seven golden cities. In Mexico, they had found the hands of the Aztecs heavy with gold and silver. They fought Montezuma for possession of this wealth, but legends whispered of greater treasure across the hills to the north. The greed of the Conquistadors led them on.

In 1536, Cabeza de Vaca, and three other members of the ill-fated expedition of Panfilo de Navarez, stumbled into Mexico City with wild tales of golden cities. Intrigued, the Spanish governor of New Spain (Mexico) sent Father Marcos de Niza to check out the story. Three years later, the good father reported that he was sure he had seen the Seven Cities of Cibola, each allegedly built of solid gold. In January 1540, Francisco Vasquez de Coronado headed north with 336 soldiers, four priests, several hundred Native American porters, and 1,500 head of horses and beef cattle.

Coronado's real footprints have long since faded away, but his footprints upon the history of the Southwest are indelible. His expedition found the Grand Canyon and opened the region to nearly three centuries of Spanish influence. Coronado never found Cibola. As near as we can tell, the legend of the Seven Cities originated with the Native American Pueblos of the upper Rio Grande Valley south of the present New Mexico-Colorado border that appeared to be golden in the late afternoon sun.

The Spanish did find that the native people living in Arizona and New Mexico were mining and using gold. Although they never built a city of gold, minerals such as gold, silver, and copper were already being worked by the Native Americans. As in Mexico, there were veins of solid metal that were sitting there for the taking. Then too, for every vein of solid gold that was found, there were stories of a vein ten times as long and wide.

The Spanish, thanks to their firearms, were able to overwhelm the Indians, take over the mines, and eventually enslave the Indians to work in them. With their better tools, the Conquistadors were able to take a great deal more gold from the mines.

The legends of lost and buried treasure in the Southwest began almost immediately, when the angry Indians began to ambush the pack trains taking the gold and silver back to Mexico. The Indians in turn hid the loot and, this time they put it where it would be hard to find.

In 1680, the Native Americans launched a massive revolt against Spanish rule that ejected the Europeans from the entire area that is now the American Southwest. The lucrative Spanish mines were closed and evidence of their existence was purposely obscured. Some of the most remote of these mines were worked by prospectors and Native Americans until relatively recent times.

Another major source of treasure were the nearly two dozen caches left behind by Jesuit missionaries when they were expelled by the Spanish government in 1767. These almost certainly exist in both Arizona and New Mexico, but they have never been recovered. Few clues exist, but there are many stories of their having been found and pilfered in the nineteenth century by prospectors who took some gold, but were unable to go back for the bulk of the treasure.

As for legends, there is probably none that has attracted more attention than that of the Lost Dutchman Mine in Arizona's portentously named Superstition Mountains, whose potential value was estimated to be roughly $100 million in the late nineteenth century, and as such could be worth $2 billion today. The mine is said to contain a gold-impregnated rose quartz vein eighteen inches thick and a second vein that is almost solid gold.

One finds a great many stories of mines in the West that have been lost and/or found by Dutchmen, but in most cases the "Dutchman" is not Dutch, but rather "Deutsche," meaning German, for there were a great many Germans among the westward immigrants of the nineteenth century. Such was the case of Jacob Waltz (aka Walz) who appeared in Arizona Territory soon after the Civil War.

In 1870, Jake Waltz and a fellow German named Jake Weiser, who he had met while working as a miner, struck out on their own and struck it rich somewhere in the Superstitions. As the legend goes, they later appeared in Phoenix, paying for drinks and dry goods with bags of nuggets.

Various stories have circulated that tell about how they found the mine. One has it that they shot and killed two Mexican miners whom they mistook for Indians and discovered they'd been digging gold. Another tale takes them to Mexico, where they rescued a man named Peralta from certain death in a saloon knife fight, and were rewarded with a look at Peralta's family treasure map. The trio are said to have taken a fortune out of the Superstitions before the Germans bought their partner's interest in the mine.

Somewhere over the years, Jake Weiser disappeared. Some say he was killed by Apaches, but others say it was Jake Waltz who did him in. For the rest of his life, Jake Waltz commuted between the mine and Phoenix, where he gave often contradictory hints to his drinking buddies about the mine's location.

The winter of 1890–1891 found an aging Jake Waltz befriending Julia Elena Thomas, an old Mexican widow who owned a small bakery in Phoenix. He told her the whole story of the mine and promised to take her there "in the spring."

Jake Waltz died before the spring of 1891, taking the secret of the mine with him. Armed with the stories and legends, two men, Sims Ely and Jim Bark, spent the next quarter century searching in vain for what they dubbed the "Lost Dutchman Mine."

Despite the fact that they are within twenty miles of the edge of the Phoenix metropolitan area, the Superstition Mountains are an almost hopeless labyrinth of dead end gullies and box canyons in which the entrance to a cave might easily be concealed against even the most exhaustive search. Bark and Ely never did find Waltz's claim, but their search helped to establish the legend.

In 1929, an American veterinarian named Erwin Ruth helped smuggle a Mexican family into the United States. These people, as the story goes, were Peraltas related to the original partner of the two Jakes. For his trouble, these Mexicans gave Erwin Ruth a copy of the original map. Ruth gave it to his father, Adolph Ruth, a federal clerk in Washington, DC.

In 1931, Adolph Ruth retired and went to Arizona to find the treasure. He hired a pair of cowboys to take him into the mountains. Several days later, Tex Barkley, the rancher for whom the cowboys worked, began to worry about the elderly city slicker out in the hills in the summer heat, and he went to look for him. Barkley found Ruth's camp, but not the man. A six-week search turned up nothing, but in December, a skull, identified as Ruth's, was found five miles from where he'd made camp. It had what looked like two bullet holes in it.

Ruth's death attracted national media attention to the Lost Dutchman Mine, and led to numerous attempts, especially during the 1950s and early 1960s, to find it. Yet it remains hidden, along with perhaps as many as 160 other treasures, in the canyons and mesas of the forty-eighth state.

Apache County

- Alec Toppington's Bear Cave Treasure: In a cave in the Carrizo Mountains in the northeastern part of the county near the New Mexico State line.

- Jim Carson's Lost Placer: Located near the Navajo village of Ganado on the Navajo Indian Reservation (The Navajo Nation).

- The Lost Shoemaker Mine: Somewhere in Northern Apache County on the Navajo Indian Reservation (The Navajo Nation).

- Major Peeple's Lost Mine: Located on Silver Creek.

- The Black River Gold: On the Black River, south of Greer in the White Mountains of the Apache-Sitgreaves National Forest.

- The Treasure of Hidden Spring: In the vicinity of Hidden or Mexican Spring, west of Rock Point Trading Post on the Navajo Indian Reservation (The Navajo Nation).

- The Lost Black Burro Mine: Somewhere in northern Apache County or northern Navajo County.

Cochise County

- Blackjack Ketchum's Bandit Treasure: Said to be hidden in Wildcat Canyon in the Chiricahua Mountains.

- Burt Alvord's Treasure: Between Willcox and Cochise, near Interstate 10 and/or US 191, possibly in the area of the Willcox Playa.

- Cochise's Treasure: The cache attributed to the great nineteenth-century Apache chief is said to be located near the area known as Cochise's Stronghold, in the Dragoon Mountains and a section of the Coronado National Forest, about 17 miles south of the town of Cochise.

- Geronimo's Lost Mine: Probably not actually worked by or even possessed by Geronimo, the mine is said to be somewhere in southern Cochise County near the Mexican border.
- The Lost Mine of the Guadalupes: Said to be in the mountains of southeastern Cochise County near the New Mexico state line.
- The Lost San Pedro Mine: Said to be somewhere in southern Cochise County.
- The Lost Shepherd's Mine: Said to be somewhere in southern Cochise County.
- The Lost Virgin of Guadalupe Mine: Somewhere in southeastern Cochise County near the New Mexico state line.
- The Apache Girl Lost Mine: In the Dos Cabezas Mountains.
- The Big House Treasure of Charleston: In or near the old mining town of Charleston in the San Pedro Riparian National Conservation Area.
- Sandstone Lost Mine: Somewhere in Cochise County.
- The Blind Man's Mine: Said to be north of Douglas, possibly in the Pendregosa Mountains.
- The Cave Canyon Mystery Treasure: Somewhere in the Huachuca Mountains.
- The Flooded Treasure of Tombstone: At the historic town of Tombstone.
- The Outlaw Brothers' Treasure: Somewhere northwest of Douglas, possibly west of US 191.
- The Train Robbery Treasure at Bisbee Junction: In the vicinity of Bisbee Junction near Bisbee.
- The Treasure Cave of Bisbee: At or near Mule Shoe Pass, about a mile west of Bisbee.
- The Treasure of Skeleton Canyon: Diamonds and gold with an original estimated value of over $2 million were lost in 1882 at Skeleton Canyon, in the Chiricahua Mountains near the intersection of Cochise, Graham, and Greenlee Counties, allegedly near a nonexistent "Davis Mountain."
- The Treasure of Fort Huachuca: On the grounds of the Fort

Huachuca Military Reservation, 20 miles south of Benson on Route 90.

- The Willcox Train Robbery Treasure: Said to have been stashed in the Dos Cabezas Mountains near Willcox.
- Cienega Benders Lost Mine: In the vicinity of Pantano, in western Cochise County or eastern Pima County.
- The Old Papago's Lost Mine: Said to be somewhere in southern Cochise County or southern Santa Cruz County, probably in the Coronado National Forest.

Coconino County

- The Coconino Lost Mine: In southwestern Coconino County, near Grand Canyon Caverns and possibly on the Hualapai Indian Reservation.
- Duncan Teller's Lost Mine: Said to be in the Elden Mountains, northeast of Flagstaff in the Coconino National Forest.
- Lemuel Dodson's Lost Gold Nugget Placer: Located in a stream or stream bed near Flagstaff.
- The Lost Monument Valley Mine: Said to be in northern Coconino County, although Monument Valley proper is at least 20 miles east of Coconino County in Navajo County.
- The Lost Padre Mine: Said to be in southwestern Coconino County, near Grand Canyon Caverns and possibly on the Hualapai Indian Reservation.
- Roy Gardner's Train Robbery Treasure: Hidden in the Flagstaff area in the Kaibab or Coconino National Forest.
- Samuel Clevenger's Treasure: About 25 miles from Lee's Ferry.
- The Bandit Treasure of Stoneman Lake: At Stoneman Lake about 35 miles south of Flagstaff.
- The Big Snake and the Bars of Gold: Hidden in the San Francisco Mountains.
- The Herman Wolf Treasure: Northeast of Canyon Diablo.
- The Lost John D. Lee Mine: In the vicinity of Vulture's Throne, in the Grand Canyon.
- The Lost Mercury Tubes Treasure: Hidden at the base of Echo Cliffs.

- The Lost Mines and Treasures of Coconino: South by southwest of Flagstaff, probably in the Tonto National Forest.
- The Lost Valise Mine: Located at or near Lava Butte.
- The Lost Waterfall Gold: At a waterfall along the Tanner Trail, near Grand Canyon.
- The Pine Springs Robber's Cache: In the vicinity of the old Pine Springs Stagecoach Station, near Mahan Mountain.
- The Thunder River Gold: In the Arizona strip area of northern Coconino County, near the Utah border.
- The Lost Mine of Sierra Azul: At Sierra Azul (Blue Mountain) in the San Francisco Peaks region, north of Flagstaff.
- The Lost Treasure of the Padres: Hidden in the San Francisco Peaks area.
- The Treasure of Roger's Lake: At Rogers Lake, southwest of Flagstaff.
- The Turquoise Shrine: Located near Inscription House on the Navajo Indian Reservation (The Navajo Nation).
- The Viet Spring Stagecoach Treasure: Hidden at Viet Spring near Flagstaff.
- The White Horse Hills Lost Mine: In the White Horse Hills.

Gila County

- The Cibicue Apaches' Lost Mine: In the Table Lands of southern Gila County.
- Dead Eye's Cave Treasure: Located near North Peak in the Mazatzal Mountains.
- The Lost Mine of the Hat Mountains: In the Hat Mountains of southern Gila County.
- The Lost Six-Shooter Mine: Somewhere in southern Gila County.
- The Lost Tonto Trail Mine: In the Tonto National Forest in northwestern Gila County.
- The Miner's Lost Mine: In the Tonto National Forest in northwestern Gila County.
- Sanders's Lost Mine: In the vicinity of Coon Creek in the Sierra Ancha Mountains.

- The Coyotero Lost Mine: In the Tonto Basin area near Mogollon Rim.
- The Lost Adams Diggings: One of many suggested possible locations is northwest of Cassadore Springs, for in the story (*see* New Mexico), Adams and his party started from the Gila Bend area (*see also* Cibola County, New Mexico).
- The Lost Gunshot Mine: On the Apache Indian Reservation.
- The Lost Mine of the Tonto Apaches: Located near old Camp Reno, in the Mazatzal Mountains.
- The Lost Silver Mountain: Somewhere on the Apache Indian Reservation.

Where Gila, Maricopa, and Yavapai Counties meet

- The Lost Gold of the Four Peaks: In the Four Peaks area, south of Payson in the Tonto National Forest.
- The Black Maverick Lost Mine: Somewhere in the Four Peaks area.
- The Lost Mine of the Two Skeletons: In the Mazatzal Mountains, near the border between Gila and Maricopa Counties.
- The Shoemaker's Lost Placer: North of the Salt River, near the Four Peaks area of the Mazatzal Mountains.

Greenlee County

- The Black Burro Lost Mine: Northwest of Clifton at the junction of the San Francisco and Blue Rivers.

La Paz County

- Pancho's Lost Gold: Located on the west slope of Little Horn Mountains southwest of Salome.
- The Lost Glory Hole Mine: Said to be in the Granite Mountains northwest of Salome.
- The Castle Dome Lost Mine: Located on a restricted military area in the Castle Dome Mountains between the old King of Arizona Mine and the town of Ehrenberg.
- John Nummel's Lost Mines: In the Trigo Mountains, north-

west of the old Castle Dome Landing on the Colorado River.

- The Lost Silver of the Trigo Mountains: In the vicinity of Clip Mountain in the Trigo Range near Cibola.
- The Lost Six-Shooter Silver Mine: Said to be south of the old freight road (roughly parallel to today's Interstate 10 and US 60) from Ehrenberg to Salome.
- The Cowboy Lost Mine: In the vicinity of Cibola near the Cibola National Wildlife Refuge on the Colorado River.
- The Crater of the Moon Lost Mine: Said to be located on Moon Mountain (possibly in neighboring Yuma County).
- The Redondo Ruins: Said to be somewhere along the Colorado River in Yuma County or La Paz County.
- The Quartzite Lost Mine: In the vicinity of Quartzite on Interstate 10.
- The Squaw's Lost Mine: Harquahala Mountains, south by east from Salome.
- The Treasure of Rancho De Los Yumas: Located on the east bank of the Colorado River, about 40 miles north of Yuma.

Maricopa County

- Aztec Treasure (aka Montezuma's Treasure): At Montezuma's Head Mountain in the Estrella Mountains, southwest of Phoenix.
- Cal Madden's Lost Mine: In the vicinity of Vulture Peak in the Vulture Mountains, south of US 60 in northwestern Maricopa County.
- The Lost Soldier Mine: Somewhere in southern Maricopa County, probably on the Luke AFB Bombing and Gunnery Range, which is a restricted area.
- Negro Ben's Lost Mine: Said to be in the vicinity of Wickenburg (which is 1 mile south of the Yavapai County).
- Swilling's Gold: Located near White Picacho Mountain.
- The Gold of Morgan City Wash: At Morgan City Wash, north of Wittman.
- The Hassayampa Flood Treasure: In the vicinity of old Fools Gulch Dam, northeast of Wickenburg (which is 1 mile south of the Yavapai County).

- The Lost Grijalva Mine: Located near Fools Canyon, where it joins Hassayampa Creek, near Wickenburg (which is 1 mile south of Yavapai County).

- The Lost Joaquín Campoy Mine: In the Estrella Mountains southwest of Phoenix.

- The Lost Mine of the Sierra Sombrera (Hat Mountain): At or near Hat Mountain, south of Gila Bend, and probably on the Luke AFB Bombing and Gunnery Range, which is a restricted area.

- The Lost Mine of Squaw Hollow: In Squaw Hollow, north of Camp Creek and northeast of Phoenix.

- The Lost Vein of the San Tans: Sand Tank Mountains on the Luke AFB Bombing and Gunnery Range, which is a restricted area.

- The Royal Treasure: Hidden northwest of Phoenix.

- The Wickenburg Stage Treasure: Said to be hidden about 9 miles west of Wickenburg.

- Treasure of Horse Thief Basin: Somewhere in the Bradshaw Mountains.

- Treasure of Walnut Grove Dam: In the vicinity of Wickenburg (which is 1 mile south of Yavapai County).

In the Superstition Mountains, which straddle Maricopa County and Pinal County

- Doc Thorne's Lost Mine: In the Superstition Mountains.

- Gonzales' Lost Mine: Located near Sombrero Peak in the Superstition Mountains.

- The Lost Bear Hunter's Mine: In the Superstition Mountains.

- The Lost Dutchman Mine: The legendary Peralta treasure discovered by "Deutsche" man Jake Waltz in the Superstition Mountains and valued at $100 million in the late nineteenth century.

- The Lost Don Miguel Peralta Mines: In the Superstition Mountains.

- The Soldier's Lost Mine: Located somewhere between Mount Ord and the Superstition Mountains.

- The Treasure of Weaver's Needle: Said to be located near the top of Weaver's Needle in the Superstition Mountains.

Mohave County

- Hiram Smith's Treasure: In vicinity of Boundary Cone near Oatman.
- Planchas De Oro: Located near Artillery Peak.
- Samuel Whitlesy's Lost Gold Ledge: South of Sitgreaves Pass, near the old ghost town of Oatman.
- The Death Trap Mountain Treasure: On or near Death Trap Mountain in the Cerbat Mountains near Topock.
- The King Tut Placer: Located on a stream or stream bed in the vicinity of Davis Mountain near Pierce Ferry, it has nothing to do with King Tutankhamen who ruled in Egypt in the fourteenth century BC.
- The Lost Nugget Placer: On a stream or stream bed southeast of Topock.
- The Shepherd Girl Lost Mine: Said to be somewhere in the Hackberry Mountains.
- The Soldier Brothers Lost Mine: Located in the Cerbat Mountains near the old mining town of Chloride.
- The Steamer Gila Treasure: In the vicinity of Crescent Spring.
- The Treasure of Fort Mohave: On the old Fort Mohave Military Reservation.
- The Treasure of Canyon Station: At the site of old Canyon Station.

Navajo County

- Henry Adams's Lost Gold Cave: Said to be northwest of a "Twin Buttes" in Navajo County, but there is a town of Twin Buttes in Pima County, so the story could refer to either or both sites.
- Henry Tice's Treasure: Located near Holbrook.
- The Lost Black Burro Mine: Somewhere on the Navajo In-

dian Reservation in northern Apache County or northern Navajo County.

- The Lost Silver of Monument Valley: In the vicinity of Merrick or Mitchell Buttes in Monument Valley on the Navajo Indian Reservation.
- Treasure of the Turquoise Shrine: At Holbrook.

Pima County

- The Arivaca Spanish Mines: Located near Arivaca Junction, south of Tucson.
- Cienega Benders Lost Mine: In the vicinity of Pantano, in western Cochise County or eastern Pima County.
- John Clark's Lost Mine: Cerro Colorado Mountains.
- Montezuma's Treasure in the Ajo Mountains: Located near the old mining town of Ajo in the mountains of the same name.
- The Bandit Treasure of Colossal Cave: Somewhere in Colossal Cave in the Rincon Mountains.
- The Black Princess Lost Mine: Said to be somewhere in the Cerro Colorado Mountains west of Amado.
- The Cursed Treasure of the Cerro Colorado: In the Cerro Colorado Mountains, northeast of Arivaca.
- The Leather Pouch and the Skeleton: The clues will lead to the treasure, somewhere in the Santa Catalina Mountains near Oracle.
- The Lost Bean Pot Placer: The clue will lead to the treasure, in the Growler Mountains west of Ajo.
- The Lost Canyon of Gold: Located in a canyon or ravine in the Catalina Mountains near Tucson.
- The Lost Clark Silver Mine: Said to be somewhere in southern Pima County, possibly on the Papago Indian Reservation.
- The Lost Esmeralda Mine: Said to be south of Tucson.
- The Lost Gold of Sonoyta: South of the old mining town of Ajo, near the United States–Mexico border.
- The Lost Jabonero Mine: Somewhere in western Pima

County, possibly near Ajo or on the Papago Indian Reservation.

- The Lost Mine of Old Guevavi: Said to be south of Tucson.
- The Lost Mine of San Cayetano: In the San Cayetano Mountains.
- The Lost Mine of the Ajo Mountains: On the Cabeza Prieta National Wildlife Range, in the Ajo Mountains of western Pima County.
- The Lost Mine of the Vampire Bats: In a bat cave in the Baboquivari Mountains. (Since vampire bats are not found in Arizona, the bats are probably Mexican freetails, and hence probably harmless to people unless rabid.)
- The Lost Mine and Treasure of Sopori: On or near the old Sopori Ranch, near Arivaca Junction.
- The Lost Mine of the Baboquivari Mountains: In the Baboquivari Mountains on or near the edge of the Papago Indian Reservation in southern Pima County.
- The Lost Mine of the Silver Stairway: In the vicinity of Gunsight Wells in the Ajo Mountains in the Cabeza Prieta National Wildlife Refuge.
- The Lost Papago Indian Placer: Located near the old mining town of Ajo, western Pima County.
- The Lost Papago Indian Arsenal: Said to be hidden somewhere in the Santa Rosa Mountains.
- The Lost Silver Ledge of the Santa Rosas: Santa Rosa Mountains east of Ajo.
- The Lost Sopari Mine: In the vicinity of Arivaca.
- The Lost Tenhachape Mine: Somewhere in southwestern Pima County, possibly on the Papago Indian Reservation.
- The Lost Treasure of Red Rock: Near Red Rock in the Silverbell Mountains, where most rocks are red.
- The Lost Treasure of the Irish Cavalier: Said to be in the Altar Valley, which runs from Pima County into the northern part of the Mexican state of Sonora.
- The Mine With the Iron Door: In the Santa Catalina Mountains near Tucson.

- The Padre Lost Mine: In the vicinity of or in the Santa Catalina Mountains northwest of Tucson.
- The Rincon Cave Treasure: Somewhere in the Rincon Mountains southeast of Tucson.
- The San Jose Del Tucson Mission Treasure: On the old mission grounds, one half mile southwest of downtown Tucson.
- The San Xavier Del Bac Mission Treasure: In the mountains surrounding the San Xavier Del Bac Mission, southwest of Tucson.
- The Treasure of Carreta Canyon: In Carreta Canyon, in the Tascosa Mountains south of Arivaca.
- The Troopers' Lost Gold: In the Quijotoa or Baboquivari Mountains, southwest of Tucson in southern Pima County.
- Henry Adams's Lost Gold Cave: Said to be northwest of a "Twin Buttes" in Navajo County, but there is a town of Twin Buttes in Pima County, so the story could refer to either or both sites.

Near the border of Pima and Santa Cruz Counties

- The Lost Mines and Treasure of Tumacacori: In the mountains surrounding the old Tumacacori Mission in Santa Cruz or Pima County.
- The Lost Blonde Mayo Mine: In the Cerro Colorado Mountains, northeast of Nogales in Santa Cruz or Pima County, possibly in the Coronado National Forest.

Pinal County

- Montezuma's Treasure at Casa Grande: In or near Casa Grande National Monument.
- The Lost Mines and Treasure of Montezuma's Head: In the vicinity of Montezuma's Head Mountain in the Estrella Range.
- Wagoner's Lost Mine: East of Weaver's Needle near La Barge Canyon.
- The Queen Creek Cache: Located on Queen Creek near Comet Peak.

- The Silver King Mine Bars: At the site of the old Silver King mine near Superior.
- Yuma's Gold: East of old Fort Grant in the Arivaipa Hills.
- The Treasure of El Tejano: Located near White Picacho Mountain east of Morristown.

Santa Cruz County

- The Lost Indian Silver Mine: Located on Carrizo Creek in the Arivaca-Tubac area of the Tumacacori Mountains.
- The Lost Opata Indian Silver: In the San Cayetano Mountains near Tubac.
- The Lost Planchas de Plata Mine: A silver mine, possibly the same as referred to in one of the previous two entries, it is said to be in southern Santa Cruz County.
- D'Estine Shepherd's Lost Mine: In the Pajarito Mountains west of Nogales.
- Hashknife Charley's Gold: Located near Sonoita, probably in the Coronado National Forest.
- "Nigger" Bense's (or Ben's) Lost Silver Mine: In the vicinity of the old Tumacacori Mission, probably near the Tumacacori National Historical Park.
- The Old Papago's Lost Mine: Said to be somewhere in southern Cochise County or southern Santa Cruz County, probably in the Coronado National Forest.
- The Lost Bells and Treasure of Guevavi: Said to have been cached south of Calabasas, near the Santa Cruz River.
- The Patagonia Hacienda Cache: Located near Patagonia.
- The Pete Kitchen Ranch Treasure: Hidden on the old Pete Kitchen Ranch, north of Nogales.
- The Treasure of Cerro Ruido: In the Pajarita Mountains west of Nogales.

Yavapai County

- Chalmer Harper's Lost Mine: Said to be in Smith Canyon.
- The Lost Pick Mine: The "clue" may indeed be harder to

find than the mine itself, said to be in the vicinity of Bronco Canyon east of Bumble Bee.

- Geronimo's Lost Gold Mine: In the Sycamore Canyon area of Prescott National Forest, near the unimproved road running between Jerome and Perkinsville.
- Lord Duppa's Lost Silver Strike: In the Bradshaw Mountains, near the town of Black Canyon.
- "Nigger" Ben's Lost Mine: Located between Congress Junction and the old McCracken mine.
- Old Mose's Lost Dutch Oven Mine: Said to be located in northeastern Yavapai County, in the Prescott National Forest.
- The Black Mesa Lost Placer: In the Black Mesa area south of the Hualapai Indian Reservation, in the northwest corner of Yavapai County.
- The Golden Cup Treasure: On Rich Hill, in the vicinity of Congress Junction.
- The Golden Nugget Placer: Located on Castle Creek north of Wittman.
- The Henry Seymour Treasure: Said to be hidden in the vicinity of Gillette.
- The Lost Creek Mine: Located south of Winslow on Clear Creek, probably nearer its headwaters in the Coconino National Forest in Yavapai County than its mouth on the Little Colorado River in Navajo County.
- The Lost Deer Head Claim: Located near Prescott, possibly in the Prescott National Forest.
- The Lost Dutchman Mine in Yavapai: Located in the Bradshaw Mountains, it is probably associated with a "Deutsche" man other than Jake Waltz of the legendary Superstition Mountains Lost Dutchman Mine.
- The Lost Ledge of the Lone Ace Desert Rat: In the vicinity of Skull Valley northwest of Prescott.
- The Organ Grinder's Lost Ledge: Located near Hillside, southwest of Prescott.
- The Silver of the Dead Apache: In the Bradshaw Mountains east of Prescott.
- The Treasure of Montezuma Well: Said to be somewhere in

the Montezuma National Monument east of Prescott.

- Yeager's Lost Gold: Said to be hidden in the vicinity of Yeager Canyon.

Yuma County

- Jesus Arroa's Lost Treasure: Said to be somewhere in the Cocopah Mountains.

- Major Peeples Lost Mine: Somewhere in western Yuma County possibly near the Colorado River, or in one of the vast military restricted areas that dominate the county.

- Old Charlie's Lost Mine: In the Tinajas Atlas Mountains on the Luke AFB Bombing and Gunnery Range Restricted Area of southwestern Yuma County.

- The Belle McKeever Lost Mine: Located somewhere near the west end of the Granite Mountains in northern Yuma County.

- The Buried Treasure of Bicuner: Located near Squaw Peak in the Laguna Mountains.

- The Frenchmen's Lost Gold Placer: On a stream or stream bed in the Eagle Tail Mountains.

- The Lincoln-Glanton Treasure: At or near an old Colorado River crossing (almost certainly not the one used by Interstate 8) near Yuma.

- The Lost Dry Wash Mine: Located on the Cabeza Prieta National Wildlife Refuge in the Sierra Pinta Mountains of southeastern Yuma County.

- The Lost Gold at Tule Tank: At Tule Tank in the Tank Mountains on the Kofa National Wildlife Refuge. (An alternate theory suggests the Tule Tanks in the Cabeza Prieta Mountains and on Cabeza Prieta National Wildlife Refuge.)

- The Lost Gold at Camel's Tank: Said to be in the Tank Mountains, northwest of Yuma and north of Wellton on what is now the US Army's Yuma Proving Ground Restricted Area. (An alternate theory suggests the Tule Tanks in the Cabeza Prieta Mountains and on Cabeza Prieta National Wildlife Refuge.)

- The Lost Indian Placer Mine: On a stream or stream bed

somewhere in western Yuma County, possibly on or near the Colorado River.

- The Lost Mine of the Orphans: Located near Tule Tanks in the Cabeza Prieta Mountains and on the Cabeza Prieta National Wildlife Refuge. (An alternate theory suggests the Tule Tank in the Tank Mountains on the Kofa National Wildlife Refuge.)
- The Lost Mission of Tinajas Altas: On the Luke AFB Bombing and Gunnery Range Restricted Area in the Tinajas Altas Mountains near the Mexican border.
- The Lost Pete Mine: Somewhere in the Gila Mountains, southeast of Yuma on the Luke AFB Bombing and Gunnery Range Restricted Area.
- The Lost Shepherd Girl Mine: Somewhere in southern Yuma County, on the Luke AFB Bombing and Gunnery Range Restricted Area or in Cabeza Prieta National Wildlife Refuge.
- The Lost Six-Shooter Gold Mine: Between Quartzite and the old mining town of Planet northeast of Yuma.
- The Lost Squaw Mine: Said to be somewhere in the Adonde Range.
- The Santa Isabel Mission Treasure: Located near the junction of the Colorado and Gila Rivers, north of Yuma.
- The Soapmaker's Lost Mine: In the vicinity of Tule Wells on the Luke AFB Bombing and Gunnery Range Restricted Area in the Tinajas Altas Mountains.
- The Treasure of San Ysidro Hacienda: In the Gila Valley, about 15 miles north of Yuma.
- The Wagon Trail Massacre Treasure: Located on or near O'Neill Pass.

NEW MEXICO

As noted above, Arizona and New Mexico can be a wonderland of light and shadow, and through this, both states have now attracted several generations of artists and art lovers, who've come here to capture and/or possess a hint of the magic that all feel in the Southwest. Santa Fe, which is a compact and picturesque state capital with just fifty thousand souls, is the second or third biggest art market in the United States, and certainly the biggest between New York and Los Angeles.

The Spanish Conquistadors came in 1540 looking not for art, but for the certainly picturesque seven cities of gold. The cities turned out to be the adobe pueblos of the Rio Grande Valley glowing in the late afternoon sun, but succeeding waves of explorers looked harder and found the Indian gold and silver mines. In 1598, Juan de Onate established a permanent colony—which he called Salinas for the nearby salt deposits—in the central part of the state, and in 1610, Santa Fe was established as a trading post. Stories about the Indian gold mines date back to this era and brim with intrigue that pitted Spanish troops against the Native people, with Franciscan friars—generally more benevolent than revisionist history paints them—caught in the middle.

New Mexico was the last of the contiguous forty-eight states admitted to the Union, and as late as the 1950s, there were Manhattan skyscrapers with more people than Santa Fe. Many visitors jet to Albuquerque, dash up Interstate 25 to Santa Fe— an hour away—immerse themselves in ''Southwest style'' with a possible day trip or ski trip to Taos, then jet home. Today, few tourists take the time to immerse themselves in the wild, remote, and intensely interesting back country reachable by the narrow, twisting side roads that jut out anonymously from two-lane blacktops like US 180 or US 285.

There are many treasures to be found on those back roads, treasures that range from gold and silver to magnificent vistas and captivating legends. One of the most amazing stories in American treasure lore is that which involves the ''Lost Adams Diggings,'' a placer so rich that gold could be picked up by the fistful. While such hyperbole is certainly not uncommon, especially in tales originating in the Southwest, the Adams site is either more widely corroborated or it is a rumor more consistently and more widely spread than most fabrications.

Adams, who like Liberace, Hammer, Cher, and Madonna, went by a single name, earned his living running freight wagons between Tucson and California. As the story goes, in 1864 near Gila Bend, an Indian raiding party burned his wagons, stole his provisions and attempted to steal his horses. He recovered his horses, but needed food. He stumbled into a mining camp, where the prospectors were willing to trade food for horses. It seems that they had also just made the acquaintance of a young Mexican man with a deformed ear who had been kidnapped and raised by Apaches, and who was now headed back to Mexico.

Apparently the young Mexican was amazed to learn that gold could buy things of "real value," such as guns and horses. When he discovered this, he told Adams and the prospectors that he knew a place where the ground was littered with gold nuggets ranging in size from that of acorns to that of "turkey eggs." He promised to take them there in exchange for a horse and gun. Naturally, an expedition was organized, made possible by Adams's horses. The Mexican led the eager Anglos across Arizona and into what is generally believed to have been the region of the Mal Pais Lava Beds in Cibola County, New Mexico.

One day, they were riding parallel to a line of high steep cliffs when the Mexican led them behind a boulder that appeared to be part of the wall. Here, they found a virtually invisible "hidden door" that led to a narrow Z-shaped canyon. The gold was present in abundance as promised. The Mexican was paid off and excused. Adams and his group then set up camp and began collecting piles of gold. As they soon discovered, a group of Apaches, including the famous Chief Nana, was camped in another part of the same canyon. Nana agreed not to bother them if they stayed away from his camp. This arrangement lasted for awhile, but eventually the greedy prospectors ventured into the area that was "off-limits." The Apaches responded by killing most of the men in the canyon, and they ambushed a party that had been sent to Fort Wingate (near present-day Grants) for supplies.

There were only three survivors. First was Emil Schaeffer, a German (called a "Dutchman") who went home from Fort Wingate with his share of gold and missed the ambush. The others were Adams himself and a man named Davidson. Having eluded the Apaches, Adams and Davidson walked to Fort Apache, Arizona, where they staggered in a week or so later, more dead than alive. Davidson did die, but Adams survived and the nuggets they had with them tended to confirm their amazing tale.

Because of the Apaches, the remoteness of the area, and Adams's poor sense of direction, no immediate attempts were made to go back to the "Zigzag Canyon." Nevertheless, the story spread and remained a part of Southwest ranch lore until it faded into obscurity in the mid-twentieth century. In the meantime, there were several attempts to find the canyon and the "Dutchman." A man named "Doc" Young tracked Schaeffer down in Heidelberg around 1900. The German confirmed the story, but insisted that he wanted nothing more to do with the diggings. There are several stories that the site was rediscovered once or twice in the 1880s or later, usually, as the stories go, by solitary men with only enough provisions to stay for a day or two, and the capacity to carry only a handful of nuggets the many miles to civilization.

There is also the tale of the man who found an inscription on an aspen tree high in the Mogollon Mountains that read: "The Adams Diggings are a shadowy naught; and they lie in the Valley of Fanciful Thought."

One of the largest and richest treasure troves currently hidden in the United States involves an enormous cache of gold bullion that was reportedly moved in from Mexico in 1933. It was a difficult time. It was the depths of the Great Depression worldwide, and an era of political upheaval in Mexico, as the tumultuous career of President (cum dictator) Plutarco Calles was reaching its climactic moment. As the story goes, gold bullion valued at $20 million in 1933 dollars was smuggled across the remote border between New Mexico and the Mexican state of Sonora, taken to northwestern New Mexico, and buried. A year later it was "stranded" when the newly elected Roosevelt Administration enacted the Gold Reserve Act of 1934. This measure, which would remain on the law books for four decades, outlawed ownership of gold by private individuals. In order to get more money into circulation during the depression, Roosevelt wanted people to turn in their gold for currency. Most did so to avoid stiff penalties. The owner— or "holder"—of the Mexican bullion did not.

As the story continues, there were attempted secret, and ultimately inconclusive, negotiations between agents of the mysterious holder and the US government in the 1950s, aimed at "selling" the gold without penalty. This trove, which could now be held legally by US citizens, may well still lie hidden, possibly somewhere in the Malpais, just upstream from the Lost Adams Diggings, or it may have long ago been spent to finance some clandestine activity.

The stories of mystery, suspicion, and superstition associated with lost treasure are often intertwined with stories of crimes and conspiracies. Few crimes in modern times sparked more suspicion and more conspiracy theories than the Watergate Scandal of the 1970s.

In 1973, during the congressional Watergate hearings, conspirator and former White House counsel John Dean casually mentioned to the Senate Committee that gadfly attorney F. Lee Bailey represented a mysterious cadre of treasure hunters that had enlisted the aid of the Nixon White House in the search for a hundred tons of gold bullion that was buried on a military reservation in New Mexico.

While the national media was preoccupied with the Watergate affair and gave the gold bullion affair little mention, the ears of treasure hunters perked up. The stumbling block was that the cache was located on military land and hence not accessible to the general public in those days at the depths of the Cold War.

The thread of the story goes back to 1937, when a group of people was deer hunting near Victorio Peak in the Hembrillo Basin, not far from the Trinity site, where the US Army would detonate the first nuclear weapon only eight years later. One of the men was a barely educated, part-Cheyenne fakir named Milton "Doc" Noss, who had done time in prison for passing himself off as a podiatrist. As the legend goes—and the legend is supported by papers that he filed with the New Mexico State Land Office—Noss discovered the opening of a cave, lowered a rope, and climbed down sixty feet with a flashlight.

In the cave, Noss first found the skeletons of twenty-seven people who'd been tied up and left to die in the cave. Upon a further search of the cave, he found Spanish armor, guns, jewelry, saddles, swords, other equipment, and a box of letters dated prior to 1880. He also found 292 ingots of gold bullion, valued in excess of $100 million.

While there is some disagreement over the original source of the treasure, it is thought that it may be the eighteenth century Felipe La Rue treasure, which was discovered and rehidden by Colonel A. J. Fountain in about 1880 and discussed in Henry I. James's 1953 book *The Curse of the San Andres*.

Through 1939, Noss undertook the difficult task of lifting eighty-eight ingots, each weighing an average of sixty pounds, from the cave. Because it was then illegal for Americans to possess gold bullion, Noss stashed the bars elsewhere. Only two other people, a pair of Mexican-American boys who briefly helped Noss, are known to have seen the cave, and no one knows where he hid the eighty-eight ingots. One of the boys is known to have died and the other disappeared. In 1949, Noss was shot to death in a disagreement with another treasure hunter named Charley Ryan.

The gold may remain hidden as much in a cloak of mystery as in rock and gravel. What can be said for certain is that the death of Doc Noss opened the door to speculation just as it closed off any knowledge of the true location of the treasure. Indeed there may be no treasure, or there may be several, if one counts the original site, plus the place or places where Doc Noss hid his eighty-eight ingots.

The speculation has spawned a myriad of tales. There are stories that Noss was investigated by the FBI for running a phony gold scam. There are stories that the US Army undertook a secret search for the gold in the 1960s or 1970s. There are stories that the Watergate conspirators hunted the gold. There are warnings that a search for the treasure would almost certainly be fatal because of all the unexploded ordnance on Victorio Peak that is left over from several decades of its being used as a gunnery range. There are even rumors that the treasure has already been found.

Any and all of these stories could be true.

Bernalillo County

- The Bell Makers' Copper: Located somewhere east of Albuquerque, possibly in the Cibola National Forest.
- Don Manuel's Lost Gold Cache: Said to be hidden southeast

of Albuquerque, possibly in the Cibola National Forest and/
or the Manzano Mountains.

- The Supper Club Lost Mine: Located near Four Hills.
- The Dolores (or Delores) Treasure: Located near the old town of San Jose.
- The Treasure of Chilili Pueblo: At or near Chilili in south-eastern Bernalillo County.

Catron County

- Captain Clooney's Lost Ledge: Said to be in Sycamore Canyon, east of Alma in the rugged Mogollon Mountains.
- The Lost Schaeffer Diggings: Said to be in the Datil Mountains.
- The Black Burro Lost Mine: In the vicinity of Apache Creek.
- The Broken Saddle Lost Mine: Located near Luna in the San Francisco Mountains.
- The Folsom Treasure: In the vicinity of Alma in the Gila National Forest.
- The Lost Waterfall Mine: Near a waterfall in the Mogollon Mountains.
- The Mexican Slave Lost Mine: Somewhere in the Mogollon Mountains.
- The Treasure Hole: Located on or near Horse Peak.
- The Wild Bunch Treasure: Thought to have been cached in the vicinity of Alma, possibly near what is now US 180.
- Turner's Lost Mine: Located somewhere along Sycamore Creek.

Cibola County

- The Lost Adams (aka Z Canyon or Zigzag Canyon) Diggings: One of many suggested possible locations, and perhaps the most likely, is at the bottom of a Z-shaped canyon in or near the Mal Pais Lava Beds south of Grants (see also Gila County, Arizona).
- The Gonzales Treasure: In the vicinity of Bluewater, about

10 miles northwest of Grants on Interstate 40 and old US 66.

- The Lost Silver Strike of Acoma: South of McCarty's on the Acoma Indian Reservation.

- The Padre Treasure of San Rafael: In or near San Rafael, about 3 miles south of Grants and adjacent to the Mal Pais Lava Beds.

- The Red Ghost Treasure: Located near McCarty's on the Acoma Indian Reservation.

- The Treasures of the Lava Beds: On or in the Mal Pais Lava Beds, southeast of Grants.

- The Twadell Treasure: In the vicinity of Grants.

- The Zuni Bells Treasure: Located near San Rafael, about 3 miles south of Grants and adjacent to the Mal Pais Lava Beds.

- Thompson's Lost Uranium: About 10 miles east of Grants near Interstate 40 and old US 66 and possibly on the Acoma Indian Reservation.

Colfax County

- Cannady's Murder Money: Hidden in Taos Canyon near the old mining town of Elizabethtown.

- Clay Allison's Red River Cache: Located near Red River Pass in western Colfax County.

- Madame Barcelo's Lost Treasure: Thought to be hidden about 40 miles east of Taos.

- The Gold Pillars Lost Mine: Located on Vermejo Creek.

- The Santa Fe Trail Treasure: In the vicinity of Raton, on or near the old route of the Santa Fe Trail (which roughly parallels US 64 in northeastern Colfax County).

- The Treasure at Palo Flechado Pass: On or near Palo Flechado Pass.

- The Wagon Train Treasure: About 25 miles east of Springer, possibly near Abbott.

Dona Ana County

- The Cave Treasure of Doc Noss/The Victorio Peak Treasure: Located in the San Andres Mountains, on or near Soledad Mountain (Victorio Peak) on the White Sands Missile Range, near the border of Dona Ana and Sierra Counties.

- The Lost Spanish Mission Treasure: Located near Mesilla, immediately south of Las Cruces.

- The Oak Springs Lost Placer: Near Canada de la Madera in the vicinity of Mesilla.

- The Old House Treasure of Amador: On the site of an "old house" at Amador near Las Cruces.

- The Treasure of Fort Seldon: At or near old Fort Seldon, which is across the Rio Grande from Radium Springs.

Eddy County

- Captain De Gavilan's Lost Gold: Said to be in the Guadalupe Mountains of Eddy County, but the mountains also reach north into Otero County and south into Texas.

- Jesse James's Lost Gold Caches: The legendary outlaw is said to have secreted caches on the Santa Clara Pueblo Reservation in Rio Arriba County and in the Guadalupe Mountains of Eddy County.

- Penasco River Treasure: Located near the Penasco River, south of Artesia.

- San Antone's Lost Mine and Treasure: Said to be in the vicinity of Dark Canyon.

- The Biting Treasure: Said to be located near Carlsbad.

- The Lost Spanish Treasure of Las Placitas: Somewhere in Eddy County.

Grant County

- Addams's Lost Mine: Located on an unidentified red mountain, north of the old mining town of Pinos Altos in the Gila National Forest, which contains many red sandstone mountains, most of them heavily forested. (This site is *probably* not to be confused with the Lost Adams Diggings in Cibola County.)

- Avery's Lost Silver Mine: On or near Bear Mountain.
- The Snively Lost Diggings: Somewhere in Grant County, possibly, but not necessarily, near Silver City.
- The Cloud Rest Treasure: Said to be hidden in the vicinity of Fierro.
- The Cooney Stage Treasure: Hidden near Cooney.
- Los Perros de la Niebla (The Dogs of the Mist) Treasure: In the mountains of the Gila Primitive Area north of Hurley.
- The Golden Giant Treasure: Located near Hurley, southeast of Silver City.
- The Lost Mine and Treasure of Fort Bayard: In or near old Fort Bayard.
- The Lost "Nigger" Diggings: Located on or near Black Mountain northeast of Pinos Altos.
- The Lost Spanish Placer: On a stream or stream bed in the Little Burro Mountains in the Gila National Forest. (Note that there is a Burro Peak in Grant County and a Little Burro Peak 100 miles east in Lincoln County and that references to these may have become scrambled over the years).
- The Swede's Treasure: Somewhere in the Little Burro Mountains. (Note that there is a Burro Peak in Grant County and a Little Burro Peak 100 miles east in Lincoln County and that references to these may have become scrambled over the years.)
- The Lost Turquoise Strike: In the vicinity of Hachita at the southern tip of Grant County in the Little Hatchet (Hachita) Mountains.
- The Spanish Cave Treasure: South of Hachita, Grant County.
- The Trooper's Lost Mine: Located near Cliff and near where US 180 crosses the Gila River.

Guadalupe County

- The Lost Spanish Mine of Anton Chico: In the vicinity of Anton Chico on Route 386 in the northwest corner of Guadalupe County near the San Miguel County line.

Hidalgo County

- The Grant House Treasure: Located in the ghost town of Shakespeare south of Lordsburg.
- The Little Dale Treasure: Said to have been hidden near Granite Gap.
- The Lost Opal Mine: In the Horseshoe Mountains near Summit.
- The Mexican Woodcutter's Treasure: In the vicinity of Dogshead Peak near Lordsburg.
- The Stagecoach Treasure of Doubtful Canyon: Located in the area known as Doubtful Canyon, near Stein's Peak.
- The Treasure of Stein's Pass: Located at or near Stein's Pass.

Lincoln County

- The Aztec Gold Cave Treasure: Said to be in a cave in the Captain Mountains north of US 380 and probably in the Lincoln National Forest.
- Hawk's Rio Bonito Silver Treasure: Along the Rio Bonito, which runs across the southern tier of Lincoln County.
- Joe Analla's Treasure: In the vicinity of White Oaks, possibly in the Lincoln National Forest.
- Mina Del Toro Treasure: In the Carrizo Hills near Carrizozo and possibly near Carrizo Peak in the Lincoln National Forest.

Luna County

- The Lost Gold: Said to be in the Florida Mountains, somewhere southeast of Deming.

McKinley County

- The Seven Cities of Cibola Treasure: Said to be somewhere in the mountains of McKinley County, although the actual "Seven Cities" are now believed to be the pueblos of neighboring Cibola and Sandoval Counties.
- The Treasure of Laguna Del Oro: Located near old Fort

Wingate, south of Interstate 40 (old Route 66).
- The Treasure of the Zuni: Located on Mesa del Toro, on the Zuni Indian Reservation in southwestern McKinley County.

Mora County

- John Fazzi's Lost Mine: Somewhere in the Rincon Mountains.
- The Grinning Skull Treasure: Said to be in the Sangre de Cristo (Blood of Christ) Mountains in Mora County, although the mountains extend south into Santa Fe County and are largely contained in Taos County to the north.
- The Lost Mine and Treasure of Spirit Springs: At Spirit Springs in the vicinity of Chacon, at the end of Route 121.
- The Three Frenchmen's Lost Mine: Located near Mora.

Otero County

- John Bellow's Lost Gold: Said to be in Alamo Canyon on the west side of Sacramento Mountains, possibly in the Lincoln National Forest.
- Mangus Colorado's Lost Gold: In the Guadalupe Mountains, which are in southeastern Otero County in an especially remote area of over 1,200 square miles that has no roads or towns.
- The Lost Mica Mine: Somewhere in Otero County.
- The Lost Mine of White Sands: In the vicinity of Tularosa, which is northeast of White Sands National Monument and east of the White Sands Missile Range.

Quay County

- The Treasure of Mesa Rica: On Mesa Rica, in the vicinity of Tucumcari.

Rio Arriba County

- Jesse James's Lost Gold Caches: The legendary outlaw is said to have secreted caches on the Santa Clara Pueblo Reservation in Rio Arriba County and in the Guadalupe Mountains of Eddy County.

- Harris Dupont's Lost Mine: In the San Pedro Mountains of southwestern Rio Arriba County.
- The Chavez Lost Copper Mine: Located in the Carson National Forest near El Rito.
- The Jealous Frenchmen Lost Mine and Treasure: Located on or near Truchas Peak.
- The Lost Spanish Queen Mine: Located somewhere in the Jemez Mountains.
- The Waterfall Lost Mine: Probably near a waterfall in the area known as Large Canyon.

Sandoval County

- Montezuma's Treasure: Said to be hidden in the vicinity of the Sandia Pueblo.
- The No-Return Cave Treasure: Located in what is probably an extremely dangerous to navigate cave in the Sandia Mountains, east of the Rio Grande.
- The Salazar Lost Mine: Said to be on Rabbit Mountain. (There is also a Rabbit Ear Mountain in Union County.)
- Stewart's Lost Mine: Located in the hills near Cabezon.
- The Little Places Treasure: In the vicinity of Placitas in southern Sandoval County.
- The Lost Hold-Walled (or Hole-Walled) Canyon: At or near Canada de Cochiti.
- The Lost Mine of Peralta Canyon: In Peralta Canyon, near Cochiti.

San Juan County

- The Arch Rock Treasure: At or near a rock formation fitting this description, between Potter Arroyo and Slane Arroyo.
- The Lost Mine and Treasure of Beautiful Mountain: At or near Beautiful Mountain in the Chuska Range.
- The Lost Navajo Rose Quartz: On the Navajo Indian Reservation (The Navajo Nation) in western San Juan County.

- The Smuggler's Gold: Said to have been hidden near Shiprock on the Navajo Indian Reservation (The Navajo Nation).
- The Treasure of the Cave of Gold: In the vicinity of Shiprock, on the Navajo Indian Reservation (The Navajo Nation).
- The Texas Outlaw Treasure of Pump Canyon: Probably located in the area of Pump Canyon.
- The Treasure of Piutche Canyon: In Piutche Canyon near Archuleta.

San Miguel County

- Jose Vaca's Cave Treasure: In a cave in the Tecolote Mountains, near the headwaters of the Pecos River, which originates in neighboring Guadalupe County.
- Lucero's Lost Cave of Gold: In a cave in the Cerro de Lauriana.
- The Pecos Pueblo Treasure: Located near Pecos in the Santa Fe National Forest and near the Santa Fe County line.
- The Spanish Church Treasure of San Augustin: In the vicinity of Laudres.
- The Treasure of Gallegos Cave: Located in Gallegos Cave near Las Vegas.

Santa Fe County

- Juan Borrego's Lost Mine: In the vicinity of Santa Fe, possibly in the Santa Fe National Forest.
- The Lost Cathedral Slope: In the San Pedro Mountains.
- The Lost Mines of Golden: Located near Golden on the Turquoise Trail (Route 14) south of Santa Fe.
- The Treasures of Los Cerrillos: In the vicinity of Los Cerrillos in western Santa Fe County.

Sierra County.

- The Cave Treasure of Doc Noss/The Victorio Peak Treasure: Located in the San Andres Mountains, on or near Soledad

Mountain (Victorio Peak) on the White Sands Missile Range, near the border of Dona Ana and Sierra Counties.

- The Padre La Rue Treasure: On Victorio Peak in the San Andreas Mountains.
- Henry Whitehead's Iron Pot Cache: In Wicks Gulch near Hillsboro.
- Russian Bill's Lost Lode: In the Black Range (Black Range Primitive Area) near the old mining town of Chloride.
- The Deer Hunter's Lost Mine: Said to be located near Hillsboro.
- The Treasure of the Caballo Mountains: In the Caballo Mountains east of the Rio Grande.

Socorro County

- The Iron Door Lost Mine and Treasure: Somewhere in Blue Canyon.
- The Rio Grande Church Treasure: Located near San Marcial, overlooking the Rio Grande.
- The Treasures of Fort Ojo Caliente: In the vicinity of old Fort Ojo Caliente southeast of Dusty.
- The Treasure of the Ladron Mountains: Ladron Mountains, about 25 miles northwest of Socorro.
- The Treasure of the Socorro Mission: At or near the Socorro Mission in the town of Socorro on US 60.

Taos County

- Gus Lawson's Lost Mine: Somewhere in the Taos Mountains near Taos.
- The Lost Mine of Juan Gallule and Techato Martinez: Somewhere on Jicarita Peak in the Taos Mountains near the San Lorenzo Pueblo.
- Padre Mora's Treasure: In or near what is now Kit Carson State Park on the outskirts of Taos.
- Simeon Turley's Lost Mine and Treasure: West of the Arroyo Hondo settlement, about 12 miles northwest of Taos.
- The Lost La Mina Perdida Mine: In the Sangre de Cristo

Mountains, near or possibly across, the Colfax County line.

- The Rio Grande Gold: Located on the Rio Grande near Taos.
- The Treasure of Tres Piedras: In the San Juan Mountains near Tres Piedras.
- White's Lost Mine: Located near Amalia.

Torrence County

- The Abo Mission Treasure: In the vicinity of the old Abo Mission on US 60.
- The Lost Sanchez Gold: At or near Manzano, possibly in the Manzano Mountains and/or the Cibola National Forest.
- The Quarai Mission Treasure: Located near site of old Quarai Mission and pueblo site near Punta de Agua.
- The Treasure of Cerro Pedernal: Located on or near Pedernal Mountain, northwest of Encino.
- The Treasure of Gran Quivira: Said to be about 6 miles southeast of Abo, although the ruins of the Gran Quivira Pueblo are 34 highway miles southeast of the Abo ruins, mainly on Route 55.

Union County

- The Tollgate Keeper's Lost Gold: About 3 miles north of Folsom, probably near the Cimarron River.
- The Treasure of Devoy Peak: On or near Devoy Peak.

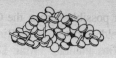

14

THE TREASURES OF CALIFORNIA

They call it the Golden State, a nickname that today implies golden sunsets on Ventura County beaches or beautiful days on Riverside County golf courses. But the name originated a year before the Golden State became a state. In 1849, California began to experience the biggest voluntary mass migration in human history, and these people weren't coming for the sunshine.

The story starts on January 24, 1848, when James Marshall, a carpenter working at John August Sutter's sawmill near Coloma in the foothills of the Sierra Nevada, found an enormous gold nugget in the waters of the American River. On second glance, there turned out to be more nuggets, and then more. It was not the first discovery of gold in California. The Spanish had found gold in the coastal hills along their El Camino Real (Royal Road) a century before, and there had been several modest, but historically significant, finds throughout California in the early 1840s. Marshall's discovery was important because there was so much gold at the original site, and because of the way his discovery fired the imagination of the nation and the world.

Hundreds of prospectors from Sacramento and San Francisco swarmed into the Sierra Nevada in 1848, but they were soon overwhelmed by the stampede of thousands, and then hundreds of thousands that poured into California the following year in what would be called the California Gold Rush of 1849.

The '49ers succeeded amazingly well. The region consisting roughly of Placer, Nevada, El Dorado, Amador, Calaveras, and Tuolumne Counties turned out to be the richest gold fields ever discovered in North America. On the whole, the take from the hills and stream beds of these and adjacent counties averaged roughly a billion dollars annually in current value for several years, and continued at some level for decades. In 1854 a miner in Calaveras County turned up a "nugget" that weighed 195 pounds. Throughout it all, prospectors and financiers dreamed of finding the "Mother Lode," the mythical, magical, monstrous vein of solid gold that had "mothered" these billions in nuggets.

Though it was never, or we should say, "not yet" found, the Mother Lode is alive and well both in the tourist advertising for the region, and in the dreams and imaginations of many who come here. There are still a few hard rock mines working successfully, and today it is hard to drive Route 49 (named for the '49ers) between Sonora and Nevada City without seeing a few people with gold pans, working patiently—and often with success, especially in the spring—in the streams that spill out of the Sierra Nevada.

The 1849 Gold Rush also brought miners to mountains and valleys of northern California. From Yuba City to the Oregon border and beyond, there were mines and diggings that yielded fortunes. While these paled by comparison to the Mother Lode, they exceeded many other, smaller gold-mining areas of the United States. The epicenter of activity was in the Klamath Mountains north of Redding, roughly consisting of what is now the Shasta–Trinity National Forest and the Whiskeytown-Shasta-Trinity National Recreation Area. Today, the latter is the only place in the system of sites managed by the National Park Service where recreational gold panning is not only permitted, but encouraged.

While the Mother Lode and Klamath country teems with tales of lost mines and stashed caches, California has other areas where the ghosts of sourdough prospectors still haunt the nights and the dark, quiet canyons. The penultimate "Gold County" in California is certainly the vast Mojave Desert area that stretches from Death Valley in Inyo County to the Mexican border, encompassing San Bernardino, Riverside, and Imperial Counties. It has had some very rich and rewarding mines through the years, but to those who are unprepared, it is a cruel, inhospitable, and potentially deadly place. While the Mother Lode is a land of wooded hillside and meadows, bisected by streams, the Mojave is dry, desolate, and often very hot. Death Valley has the hottest recorded temperatures in the Western Hemisphere. They don't call it "Death" Valley for naught. The sands still blow across the bleached bones of travellers and prospectors whose luck ran out here.

Of the "lost" and abandoned mines in the area, few have fired imaginations more than the Breyfogle Mine. The story began in 1864, when a man named Jake Breyfogle staggered into an Inyo County trading post shoeless, with his clothes in shreds and a sunburn on his bald head so severe that people thought he'd been scalped. Nevertheless, his pockets were filled with the richest gold ore anyone had seen. He had been lost in Death Valley for many days and it took him at least that long to recover his senses. When he did, he told his tale of finding the gold in what were probably the Funeral Mountains, but he couldn't remember the exact location. For over a century, there have been many attempts—throughout Death Valley—to find the Lost Breyfogle Mine. Indeed there have been *so* many, that the term "breyfogling" became synonymous with looking for lost mines.

Few states offer a wider variety of terrain and environments for the would-be breyfogler than California. It is a vast and diverse state. The longest straight-line distance in a single state anywhere in the continental United States is in California. It is farther from the northwest corner of the Golden State at Smith River to the state's southeast corner at Winterhaven than it is from New York City to Saint Louis or Jacksonville. In that distance, one goes from redwood rain forests on the rugged Pacific shoreline to the rolling sand dunes of vast deserts. Within this area have been some of the richest gold strikes in human history, and some of the most intriguing tales of fortunes lost, fortunes found, and fortunes mislaid with a nod and a wink.

Alameda County

- The Treasure of the Higuena Ranch: Said to be on the site of the old Rancho de Los Tularcitos.

Alpine County

- Snowshoe Thompson's Lost Mine: In the Diamond Valley area of the Sierra Nevada near, or possibly across the El Dorado County line.

Amador County

- Joe Williams's Lost Treasure: Somewhere along Dry Creek near Drytown on Route 49.
- The Lost Gold Bars From the Fremont Mine: In the vicinity of Amador City on Route 49.
- The Lost Mine of Dry Creek: Located on Dry Creek near Volcano.
- The Lost Mine of Rancheria Creek: Located on Rancheria Creek near Drytown.
- The Spanish Miner's Treasure: Located in the hills north of Fiddletown.

Butte County

- The Lone Indian's Lost Ledge: In the vicinity of Oroville.

Calaveras County

- Bill Moyle's Lost Mine: Said to be west of Stanislaus River, near Melones, possibly inundated by the waters behind one of the dams in the area.
- Charles Prickett's Lost Placer: Located near the western base of Bear Mountain.
- Joaquín Murietta's Buried Treasure: The notorious bandito is said to have buried his plunder throughout Calaveras and adjacent counties, especially in the vicinity of the county seat at San Andreas.

- Buster's Lost Gold: Said to be hidden in or near the old San Antonio mining camp.

- Negro Jim's Lost Treasure: Hidden in or near the site of At Campo Seco.

- The Broken Rifle Lost Ledge: Somewhere in the Blue Mountains.

- The Gunbarrel Lost Mine: Indicated as being in the Blue Mountains, the story may be a variation on that of the previous entry.

- The French Miner's Treasure of Chile Gulch: Said to be cached at Chile Gulch, 2 miles south of Mokelumne Hill on Route 49.

- The Lost Ledge of Sheep Ranch: Located in the rugged hill country between Sheep Ranch and Douglas Flat.

- The Mexican's Treasure at Whiskey Slide: At or near the old mining camp of Whiskey Slide, northeast of San Andreas.

- The Red Handkerchief Lost Mine: Though the clue and namesake have probably been gone for over a century, the mine itself still awaits discovery somewhere on the Stanislaus River, near a place once called "Big Tree" that is probably the same as, or contained within, today's Calaveras Big Trees State Park.

- William Miner's Treasures: Said to have been hidden near Angels Camp in Calaveras County and near Woodland in Yolo County.

- Joaquín Murietta's Treasures: During the 1850s, California's notorious bandit and murderer was active in and around nearly all the mining camps and towns in Calaveras, Fresno, Mariposa, San Diego, San Joaquin, Shasta, Tuolumne, and Ventura counties, and caches of his plunder are rumored to be hidden in each.

Colusa County

- Charles B. Sterling's Lost Cache: Said to be hidden in the vicinity of Fresno Crossing, probably on the Sacramento River.

Contra Costa County

- John Marsh's Lost Treasure: Located on Marsh Creek, on the east side of Mount Diablo near Brentwood in the Sacramento Delta country of east Contra Costa County.
- The Spanish Padre's Lost Mine: Located somewhere on Mount Diablo.

Del Norte County

- The Burned Cabin Lost Mine and Treasure: In Redwood National Park in the vicinity of Requa, at mouth of Klamath River overlooking the Pacific Ocean.

El Dorado County

- Edmund Cooper's Lost Treasure: Located near Diamond Springs southwest of Placerville.
- John Chapman's Treasure: Said to be hidden at Alabama Flat near Placerville.
- Snowshoe Thompson's Lost Mine: In the Diamond Valley area of the Sierra Nevada near, or possibly across the Alpine County line.
- The Chinese "Tong Man" Treasure: Said to have been hidden in the vicinity of Placerville.
- The Chinese Treasure of Volcanoville: In or near old mining town of Volcanoville (not to be confused with the town of Volcano in neighboring Amador County) near the Rubicon River at the end of an unimproved mountain road.
- The Gold Ore of Twin Lakes: Near Twin Lakes, northwest of Placerville.
- The Golden Cache of Lotus: Said to be located near Lotus, between Auburn and Placerville, possibly near Interstate 80.
- The Lost Sonora Ledge Mine: Said to be somewhere in El Dorado County, although the town of Sonora is located in nearby Tuolumne County.
- The Lost Treasure of Hangtown: Hidden somewhere just outside Placerville, which was known as "Hangtown" in the nineteenth century.

- The Oregonian's Lost Ledge: Reported to be north of Georgetown near the Placer County line and possibly in the Folsom State Recreation Area.

- The Pack Rat's Gold Slugs: Rumored to be located at the old Forty Mile House near Shingle Springs.

Fresno County

- Schippe's Lost Gold: Hidden somewhere in the vicinity of Horseshoe Bend on the south fork of King's River.

- Tiburcio Vasquez's Treasure: Located near the old settlement of Las Juntas, at the confluence of the San Joaquin River and Fresno Slough.

- The Lost Indian Ledge: Said to be located in the San Joaquin Valley in Fresno County, an area so vast as to require one to research local lore in more detail before attempting to find such a place.

- Joaquín Murietta's Treasures: During the 1850s, California's notorious bandit and murderer was active in and around nearly all the mining camps and towns in Calaveras, Fresno, Mariposa, San Diego, San Joaquin, Shasta, Tuolumne, and Ventura counties, and caches of his plunder are rumored to be hidden in each.

Glenn County

- The Bentz Company Robbery Treasure: Located near Biggs, which is on a back road near Route 99.

Humboldt County

- The Treasure of the San Francisco Mint: Cached somewhere at or near Shelter Cove, near Point Delgado in the Benbow Lake State Recreation Area.

- The Lost Wire Gold of Willow Creek: Said to be located at Willow Creek near Eureka.

Imperial County

- Captain Arroa's Treasure: Said to have been hidden near Superstition Mountain, possibly in a restricted area of a US Navy Gunnery Range.

- Charles Knowles's Lost Mine: Located in the vicinity of Carrizo Wash near Carrizo Mountain, possibly in the restricted area of the Carrizo Impact Area.

- Ebner's Lost Copper Ledge: Located near Mammoth Wash in the Chocolate Mountains, which are now designated as the Chocolate Mountains Impact Area and as such are closed to public access.

- Harpendings Lost Silver Ledge: North of Glamis in the Chocolate Mountains, which are now designated as the Chocolate Mountains Impact Area and as such are closed to public access.

- Henri Brandt's Lost Mine: West of Brawley on or near Superstition Mountain, possibly in a restricted area of a US Navy Gunnery Range.

- Pegleg Smith's Lost Black Nugget Placer: Said to be in a stream or stream bed that is within a 30-mile radius of the Salton Sea, which also extends into Riverside County.

- Sullivan's Lost Ledge of Gold: Somewhere in the Cargo Muchacho Mountains.

- The Badger Hole Treasure: In the vicinity of the shrine of E-Vee-Taw-Ash, north of Winterhaven.

- The Carrizo Stage Station Treasure: Somewhere east of the old Carrizo Stage Station.

- The Dutchman's Lost Laguna Gold Ledge: South of the old mining town of Picacho in the Chocolate Mountains, which are now designated as the Chocolate Mountains Impact Area and as such are closed to public access.

- The Lost Gold of the Algodones Dunes: In Algodones sand dunes, southwest of Ogilvy.

- The Lost Mule Shoe Gold: Said to be west of the old mining town of Picacho in the Chocolate Mountains, which are now designated as the Chocolate Mountains Impact Area and as such are closed to public access.

- The Lost Ships of the Desert: Grounded ships reported in the Salton Sink area.
- The Rodriques Lost Ledge: Located near the old mining town of Picacho in the Chocolate Mountains, which are now designated as the Chocolate Mountains Impact Area and as such are closed to public access.
- The Black Butte Lost Mine: Somewhere in the Chocolate Mountains, which are now designated as the Chocolate Mountains Impact Area and as such are closed to public access.
- The Spanish Treasure of Harper's Well: In the vicinity of Harper's Well, near where San Felipe Creek and Carrizo Wash meet.
- The Spanish Treasure of Carrizo Wash: Located on Carrizo Wash, near Harper's Well, this is possibly not a variation on the story of the previous entry.
- The Treasure at Travertine Rock: Located at Travertine Rock.
- The Treasure of Signal Mountains: Located north of Signal Mountain.
- The Treasure Ship of the Desert: At Kane Springs.

Inyo County

- Thomas Shannon's Treasure: In the vicinity of Emigrant Spring in Death Valley National Monument.
- The Lost Gold of Charlie's Butte: In the vicinity of Fish Springs in the Owens Valley.
- The Bullion Ship of Owens Lake: Located at Owens Dry Lake, west of Death Valley National Monument.
- Death Valley's Burned Wagons Treasure: Located near Death Valley Junction, about 15 miles east of Death Valley National Monument.
- Cornish's Lost Gold Nuggets: Northeast of the old mining town of Allegheny.
- The Treasure Canyon of the Coso Ancients: Coso Mountains, Northwest of Trona.
- Albert Coe's Lost Mine: Said to be somewhere in the Pan-

amint Mountains that run for over 100 miles along the west side of the Death Valley National Monument.

- Albert M. Johnson's Gold Coin Cache: Said to be hidden near the "Race Track" about 25 miles south of Scotty's Castle at the north end of Death Valley National Monument.
- Alec Ramey's Lost Bonanza: North of Ubehebe Peak in Death Valley National Monument.
- Bob Black's Lost Quartz Float: Located on the southern end of the Avawatz Mountains in Death Valley National Monument.
- Charles Wilson's Chicken Bone Gold: In the vicinity of Surprise Canyon, in the Panamint Mountains west of Death Valley National Monument.
- Death Valley Scotty's Lost Mine: Near Scotty's Castle at the north end of Death Valley National Monument.
- Gomer Richard's Lost Ledge: In the vicinity of Boundary Canyon in Death Valley National Monument.
- Jack Stewart's Lost Ledge: In the Panamint Mountains, between Scotty's Castle and Stovepipe Wells in Death Valley National Monument.
- John Goller's (or Gollier's) Lost Placer: In a stream bed in the Panamint Mountains in Death Valley National Monument.
- Juan Reynas's Lost Placer: In Colorado Canyon near Ubegebe Mountain in Death Valley National Monument.
- Shorty Harris's Lost Mine: In Emigrant Canyon near Skidoo in Death Valley National Monument.
- The Chinaman's Lost Ledge: In the vicinity of Six Springs Canyon in Death Valley National Monument.
- The Golden Eagle Lost Mine: Located near Furnace Creek in the center of Death Valley National Monument.
- The Gunsight Lost Mine: In the Funeral Mountains of Death Valley National Monument.
- The Lost Arch Mine: Said to be somewhere in Death Valley National Monument, probably in the valley itself.
- The Lost Breyfogle Mine: In the Funeral Mountains of Death Valley National Monument.

- The Lost Gold Ledge of Manly Peak: Located on Manly Peak in the Panamint Mountains of Death Valley National Monument.

- The Lost Gold of Paiute Cave: Located in a cave on Telescope Peak in the Funeral Mountains of Death Valley National Monument.

- The Lost Gold of Hungry Hill: In Death Valley National Monument, north of Wingate Pass.

- The Lost Mine of Colorado Canyon: In Death Valley National Monument area.

- The Lost Octagonal Coins of the Panamint Mountains: A cache of eight-sided, probably Spanish or Mexican, coins is hidden somewhere in the Panamint Mountains of Death Valley National Monument.

- The Madman's Lost Bonanza: Located near Ubehebe Crater at the north end of Death Valley National Monument.

- The Russell-Huhn Lost Ledge: Reported to be in the Panamint Mountains in Death Valley National Monument.

Kern County

- Finley's Lost Ledge: Located on Barbarossa Ridge in the Piute Mountains.

- The Lost Nuggets of John Goller (or Gollier): In the El Paso Mountains east of Mojave.

- The Lost Carry Mine of Randsburg: In the Argus Mountains near Randsburg.

- Old Schippe's Lost Mine: The legend states that the mine is "somewhere in Kern County," but the county is larger than Rhode Island, so local inquiry is not only recommended, but essential, to narrowing down the specifics.

- Samuel Holmes's Lost Mine: Located near the old mining town of Havilah.

- Slim Winslow's Lost Mine: In the Piute Mountains.

- The Coin Treasure of Isabella: In the hills near Isabella.

- The Indian's Lost Placer of Last Chance Canyon: In the El Paso Mountains, northwest of the old ghost town of Garlock.

- The Lost Gold Coins in Horse Canyon: In Horse Canyon near Tehachapi.
- The Lost Gold Placer of Red Rock Canyon: In a stream or stream bed in Red Rock Canyon, east of Mojave.
- The Lost Mine of Frazier Park: Near Frazier Park in the Tehachapi Mountains, possibly in the Los Padres National Forest and possibly near or across the Ventura County line.
- The Lost Mine of the Havilah Mountains: Located north of Lightning Peak in the Havilah Mountains.
- The Lost Padre Mine: In San Emigdio Canyon in the Tehachapi Mountains.
- The Lost Silver and Gold of Lerdo: Said to be located at Lerdo, near Bakersfield.
- The Lost Yarborough Mine: Located on the Kern River in the vicinity of Kernville, probably in the Sequoia National Forest, and possibly near or across the Tulare County line.
- The Mine Bullion Treasure at Robbers Roost: Said to have been cached at Robber's Roost, near Freeman Junction on Route 14 north of Red Rock Canyon.
- The Stagecoach Treasure at Freeman Junction: Said to be near Freeman Junction, possibly also at Robber's Roost, and possibly it is a variation of the treasure story referred to in the preceding entry.
- The Saloonkeeper's Treasure: Specifically rumored to be hidden at junction of Greenhorn Gulch and Freeman Gulch.
- The Spanish Carreta Treasure: Located near Boron on the old Death Valley-Mojave Road.
- The Treasure of Cache Creek: Said to be in Cache Creek Canyon, near Tehachapi Pass on Route 58.
- The Treasures of Greenhorn Gulch: Said to be in Greenhorn Gulch near Oak Creek in the Greenhorn Mountains.
- The Washwoman Lost Mine: Said to be located on Bodfish Creek.

Lake County

- The Mean Woman Treasure: Said to have been stashed, possibly by the "Mean Woman" herself, in the vicinity of the old Bradford Mine.

Lassen County

- Henry Gordier's Lost Treasure: Said to be in the vicinity of Janesville, which is about a mile west of US 395.
- The Lost Treasure of Bloody Springs: Located at Bloody Springs, which is near the Pit River, southeast of the Shasta County town of Pittville.
- Brockman's Lost Ledge: Somewhere southwest of Horse Lake.

Los Angeles

- The Malabar School Treasure: Located on the Malabar School grounds in the city of Los Angeles.
- The Spanish Treasure of Elysian Park: Said to be somewhere in Elysian Park in the city of Los Angeles.
- The Treasure of Cahuenga Pass: In or near the Hollywood Bowl in the city of Los Angeles near the Hollywood Freeway (US 101).
- Alessandro Repetto's Treasure: At King's Hills, near Monterey Park, east of downtown Los Angeles on the San Bernardino Freeway (Interstate 10).
- The Treasure of the San Gabriel Mission: In or near the hills around Monterey Park.
- Andreas Duarte's Treasure: Located near Duarte, east of Pasadena on Interstate 210.
- The McNally Ranch Lost Treasure: In Norwalk, southeast of downtown Los Angeles on the Santa Ana Freeway (Interstate 5).
- Miguel Leonis's Lost Treasure: In or near Calabasas, northwest of downtown Los Angeles on the Ventura Freeway (US 101).
- The Bandit Treasure of Camp Oak Grove: At or near Camp Oak Grove in the San Gabriel Mountains and the Angeles National Forest, northeast of the city of Los Angeles.
- The Buried Treasure of Palmdale: North of Acton Junction near the old road to Palmdale, which is now superseded by Route 14.

- The Convict's Lost Treasure: In or near the old Governor Mine, which is in the vicinity of Palmdale.
- The Indian Squaw's Lost Placer: At the mouth of Fish Canyon and San Gabriel Wash, near Azusa, east of Pasadena on Interstate 210.
- The Lost Mine of Mount Disappointment: Near Mount Disappointment in southern Los Angeles County.
- The Mexican's Lost Treasure: Reportedly hidden in or near the present city of Chatsworth in the northwestern corner of the San Fernando Valley near the Ventura County Line.
- The Padre's Lost Mines: Reported to be located in various places in Los Angeles County, including on Gleason Mountain, on Saw Mill Mountain, and near Big Tujunga Wash.
- The Pirate Treasure of Duncan Ranch: Said to be located in the Duncan Ranch area of Manhattan Beach, southwest of Los Angeles on Route 1 (the Pacific Coast Highway).
- Tiburcio Tapia's Treasure: Located on the old Cucamonga Ranch near Pomona on the San Bernardino County line.
- The Treasure of San Francisquito Canyon: In San Francisquito Canyon on the old San Francisco land grant, north of Saugus in the San Fernando Valley.
- The Treasure of Santa Susana Pass: Hidden somewhere on Santa Susana Pass between Santa Susana and the San Fernando Valley.

Madera County

- James D. Savage's Lost Gold: Located on Fresno Creek near Coarsegold.

Marin County

- The Lost Treasure of Red Rock Island: Said to be located on Red Rock Island south of the mouth of San Rafael Creek, although there is also Red Rock Island a few miles away at the point where Marin County, San Francisco County and Contra Costa County meet, about 100 yards south of the Richmond–San Rafael Bridge.

- Sir Francis Drake's California Treasure: The English buccaneer and rascal is said to have hidden loot at Drake's Bay in the Point Reyes National Seashore, where he landed and camped in 1579.

Mariposa County

- The Cave Escondido Treasure: Located on Merced River near Bagby.
- The Dying Mexican's Lost Ledge: Located near Bagby.
- The Lost Bean Pot Treasure: Located on Maxwell Creek near the old mining town of French Mills.
- The Mariposa Tax Collector's Treasure: Said to be hidden on Deadman's Creek near Agua Fria.
- Joaquín Murietta's Treasures: During the 1850s, California's notorious bandit and murderer was active in and around nearly all the mining camps and towns in Calaveras, Fresno, Mariposa, San Diego, San Joaquin, Shasta, Tuolumne, and Ventura counties, and caches of his plunder are rumored to be hidden in each.

Mendocino County

- The Safe Treasure of Mendocino: Said to be hidden north of Mendocino, possibly near Route 1, or in the rugged coastal hills.

Merced County

- The Sonora Stage Treasure: Said to have been hidden somewhere in the vicinity of Snelling on the Merced River.
- The Basque Brothers' Treasure: Cached in Six-Mile Canyon near Loyalton.
- The Lost Miner's Treasure of Snelling: Located near Snelling.
- The Treasure of Rancho Centinella: In vicinity of Rancho Centinella.

Modoc County

- Captain Holden Dick's Lost Mine: Located between Pine Valley and Oak Creek in the Warner Mountains.
- Dan Hoag's Lost Treasure: Hidden in Surprise Valley near old Fort Bidwell possibly in the Modoc National Forest or on the Fort Bidwell Indian Reservation.

Mono County

- Shepard's Lost Mine: Located near the north end of Mono Lake, possibly near Route 167.
- The Bodie Stage Treasure: Located on the stage route, north of the old ghost town of Bodie.
- The Irish Gambler's Treasure: Hidden in the ghost town of Bodie, now designated as Bodie State Historic Park.
- The Lost Cement Mine: In the Sierra Nevada near the headwaters of the Owens River.
- The Lost Gold of McGee Creek: Located on McGee Creek near McGee Lake.
- The Soldier's Lost Mine: In the vicinity of Bridgeport on US 395.
- The Spanish Mine of Sonora Pass: Located near the summit of Sonora Pass on Route 108 near the Sonora County line.

Monterey County

- Sir Francis Drake's Treasure: The English explorer and pirate may have stashed a cache on the coast near Monterey when he passed this way en route to his legendary landfall at Drake's Bay in Marin County in 1579.
- The Chinese Buried Gold: Said to be hidden in the vicinity of the farming community of Gonzales on US 101 south of Salinas.
- The Lost Indian Gold of Carmel Mission: Located near the mouth of the Carmel River near Point Lobos.
- The Lost Mine of Ventana Park: In Ventana Park southwest of Monterey.

- The Sanchez Treasure: Reported to be hidden in or around the city of Monterey.

On or near the Monterey San Benito County line

- Tiburcio Vasquez's Treasure: Hidden in or near Pinnacles National Monument, which is on the border between San Benito and Monterey.

Napa County

- The Napa Stagecoach Treasure: Located on the "old stagecoach route" north of Napa, which correlates today to either Route 29 or the Silverado Trail.
- The Quicksilver Payroll Cache: Hidden on Pope Creek in Pope Valley.

Nevada County

- Casserly's Lost Treasure: Said to be hidden somewhere in Nevada City on Route 20.
- Jefferson Casserly's Treasure: Buried along Deer Creek.
- Mayberry's Lost Treasure: In the vicinity of the old mining camp of Bloody Run northwest of Nevada City.
- The Lost Padre Mine: Said to be in the vicinity of Nevada City.
- The Miner's Treasure at Lake Vera Lodge: Hidden at the Lake Vera Lodge.

Orange County

- The Lost Mines and Treasure of the San Juan Capistrano Mission: In or near the Capistrano Mission, and/or in the coastal hills beyond.
- The Treasure of the Irvine Ranch: On the grounds of what was the old Irvine Ranch near Corona del Mar.

Placer County

- Nelson McCormick's Lost Gold: Located near the old mining town of Forest Hill.
- The Pack Rat's Treasure: Hidden somewhere along the American River southeast of Gold Run.
- The Treasure of the Donner Party: Said to have been buried near Truckee in 1846 by the ill-fated group who were caught in heavy snow during an attempt to cross the Sierra Nevada in mid-winter.
- Yankee Jim's Lost Mine: In the vicinity of the old mining camp of Yankee Jim.

Plumas County

- Henry Gorder's Lost Treasure: Located somewhere on Baxter Creek.
- Lingard's Lost Lake of Gold: Reported to be somewhere in Plumas County, although the identity of the specific lake is uncertain.
- Marks's Lost Placer: Near Rich Bar, in a stream or stream bed in the vicinity of Camel Peak.
- The Lost Ledge of the Brown Hills: In the Brown Hills district, near the village of La Porte.
- The White Mule Lost Mine: Located near Taylorsville in the Plumas National Forest.

Riverside County

- Figtree John's Lost Placer: On a stream or stream bed in the Santa Rosa Mountains.
- The Bandit Treasure of Indian Wells: Hidden near Indian Wells between Palm Desert and Indio on Route 111.
- The Frenchman's Bull Ring Mine: In the Black Hills near Wiley's Well and Mule Springs.
- The Hungarian Lost Mine: In or near Joshua Tree National Monument, near the border between San Bernardino and Riverside counties.

- The Lost Gold at Tabesca Tank: Located near Tabesca Tank between the Orocopia and Chocolate Mountains. (Note that the Chocolate Mountains are now designated as the Chocolate Mountains Impact Area, and as such are closed to public access.)

- The Lost Gold of the Hexie Mountains: In the Hexie Mountains, northeast of Indio, possibly in or near Joshua Tree National Monument, near the border between San Bernardino and Riverside counties.

- The Lost Mine of the Little San Bernardinos: Said to be in the Little San Bernardino Mountains south of Joshua Tree National Monument.

- The Lost Papuan (or Papago) Diggings: Said to be in the McCoy Mountains northwest of Blythe.

- The Lost Treasure of Dos Palmas: In the vicinity of Dos Palmas near Mecca.

- The San Franciscan's Lost Mine: In the Cottonwood Mountains near Cottonwood Springs.

- Juan Chavez's Lost Treasures: Two different treasures have been attributed to him, one near Sage in Riverside County, and one on Agua Blanca Creek, near its confluence with Piru Creek in Ventura County.

- The Lost Schwartz Diggings: In the vicinity of Rockhouse Canyon in the Santa Rosa Mountains.

- The Santa Rosa Lost Indian Emerald Mine: Located in or near Rockhouse Canyon in the Santa Rosa Mountains.

On or near the border of Riverside and San Diego Counties

- The Lost Gold Pockets of the Santa Rosa Mountains: In the Santa Rosa Mountains which cross the border of the two counties north of Borrego Springs, and which are largely contained within the Anza-Borrego State Park and the Santa Rosa Mountains Scenic Area.

- The Lost Emerald Mine: Somewhere in the Santa Rosa Mountains.

- The Innkeeper's Treasure: Between the old Warner Ranch and Aguanga.

San Benito County

- The Lost Silver Mine of San Carlos Borromeo Mission: Located on Gabilan Peak, near Hollister, San Benito County.

San Bernardino County

- Alvord's Lost Mine: Somewhere in the Alvord Mountains.
- Bernard McFadden's Treasure: In or near the Lake Arrowhead Hotel at Lake Arrowhead.
- Big Stoop's Lost Nuggets: In the vicinity of Cedar Canyon on the road that runs past the old Death Valley mine.
- Butler's Lost Mine: Located near the Coyote Mountains in the northern part of the Anza Borrego Desert.
- Buzztail Treasure: Said to be hidden somewhere north of Barstow near Lane Mountain.
- Camilo Ynita's Treasure: Located on Mount Burdell.
- Edward Shaw's Lost Gold Vein: Somewhere in the Bullion Mountains near Daggett. (Note that much of the Bullion Mountain Range is located within the Twenty-nine Palms Marine Corps Training Center Restricted Area, and as such there is no public access.)
- Free Gold: A tantalizing prospect somewhere in the Providence Mountains near Goffs, possibly in or near the Providence Mountains State Recreation Area, which is contained within the East Mojave National Scenic Area.
- Hermit John's Lost Mine: Somewhere in the Sheep Hole Mountains, northeast of Dale Lake and north of Route 62. (Note that part of the Sheep Hole Mountain Range is located within the Twenty-nine Palms Marine Corps Training Center Restricted Area, and as such there is no public access.)
- Jamison's Lost Tub Placer: On a stream or stream bed in the Turtle Mountains north of Route 62 and west of US 95.
- Jim Dollar's Lost Gold: In the Turtle Mountains north of Route 62 and west of US 95.

- John McCloskey's Lost Ledge: In the vicinity of Salt Springs, about 30 miles north of Baker.

- Johnny Lang's Lost Treasure: Located near the Lost Horse Mine in Joshua Tree National Monument, which spans the border between San Bernardino County and Riverside County.

- Lizard Bill's Lost Silver: Somewhere in the Piute Mountains, probably south of Interstate 40.

- The Lost Amboy Mine: Said to be in the vicinity of Amboy, which is located on the old National Trail Highway, about 11 miles south of Interstate 40 and near Amboy Crater and Bristol Dry Lake.

- The Gold of Amboy Crater: Located near Amboy Crater, and possibly related to the previous entry.

- The Lost Black Slate Gold of Cucamonga: About 4 miles south of Cucamonga Peak.

- The Lost Spanish Gold of the Paiute (Piute) Mountains: Said to be in the Piute Mountains near Lanfair.

- Louis Rubidoux's Treasure: Located on the south slope of Stover Mountain south of Colton near the Riverside County line.

- Old Charlie's Lost Prospect: Said to have been in the Turtle Mountains north of Carson Wells.

- Old Man Lee's Lost Lode: Somewhere in the Bullion Mountains (Note that much of the Bullion Mountain Range is located within the Twenty-nine Palms Marine Corps Training Center Restricted Area, and as such there is no public access.)

- Pack Rat Joe's Lost Mine: In the Turtle Mountains near Carson Wells.

- Pat Hogan's Lost Gold Cache: Hidden in or near the old mining town of Calico east of Barstow and about 5 miles north of Interstate 15.

- The Hidden Treasure of Gifford Gardens: Said to be in the town of Calico, and possibly a variation on the previous entry.

- The Rattlesnake Canyon Gold: In the vicinity of the junction of Rattlesnake Canyon with Burns Canyon, northeast of Yucca Valley near Route 247.

- The Golden Cavern of Kokoweef Peak: A cave located somewhere on Kokoweef Peak in the Ivanpah Mountains.

- The Lost Arch Placer: On a stream or stream bed in the Turtle Mountains, between Goffs and Rice.

- The Lost Bismuth Lode: Said to be located in what is now the East Mojave National Scenic Area, north of Beecher's Springs, near Kelso.

- The Lost Burro Ledge: In the Kelso Mountains of the East Mojave National Scenic Area, north of Baghdad. (Note that there is a Burro Peak in Grant County and a Little Burro Peak 100 miles east in Lincoln County and that references to these may have become scrambled over the years.)

- The Lost Diamond Mine of San Bernardino: Said to be in the vicinity of the city of San Bernardino.

- The Lost Dutch Oven Mine: Somewhere in the Clipper Mountains between Interstate 40 and the old National Trail Highway.

- The Lost Gold Brick of Barstow: Said to be along the Mojave River near Barstow.

- The Lost Iron Door Mine: Presumably an actual iron door leads to this mine on Table Mountain in the Providence Range in the East Mojave National Scenic Area.

- The Lost Ledge of the Sheep Hole Mountains: Said to be somewhere in the Sheep Hole Mountains, southeast of Amboy.

- The Lost Mine of Indian Gulch: At the place known as Indian Gulch in the Ivanpah Mountains.

- The Lost Mine of Rattlesnake Canyon: Said to be in Rattlesnake Canyon, a gulch presumably infested with such creatures, in or near the city of San Bernardino.

- The Lost Mine of the Bullion Mountains: The site is reported to be north of the town of Twenty-nine Palms, which would place it in that part of the Bullion Mountain Range which is located within the Twenty-nine Palms Marine Corps Training Center Restricted Area, and as such there is no public access.

- The Lost Mine of Whipple Wash: Said to be in Whipple Wash, about 14 miles north of Parker Dam on the Colorado River.

- The Lost Missouri Mine: Located in the Cady Mountains, northeast of Daggett.

- The Lost Placer of Minnelusa Canyon: On a stream or stream bed in Minnelusa Canyon, north of Big Bear Lake in the San Bernardino National Forest.

- The Lost Quail Lode: Popular legend suggests that it is located somewhere in the Providence Mountains north of Barstow, although the Providence Mountains are mostly east of Barstow, and there is a Quail Range of Mountains north of Barstow that is largely contained in the restricted areas of the Naval Weapons Center and the US Army's Fort Irwin.

- The Lost Shotgun Mine: Said to be in the Sheep Hole Mountains east of Dale Dry Lake.

- The Lost Silver of Granite Mountain: Said to be in the Granite Mountain near Old Woman Springs, although the Granite Mountains are largely contained in Riverside County and there are Old Woman Mountains to the north in San Bernardino County.

- The Lost Silver of Homer Mountain: Somewhere on Homer Mountain near Bannock.

- The Lost Spanish Mine of Yucaipa: Located near Yucaipa on the Riverside County line.

- The Lost Van Duzen Mine: Located in Van Duzen Canyon in the San Bernardino Mountains and the San Bernardino National Forest near Big Bear Lake.

- The Needles Bank Robbery Cache: Hidden in the vicinity of Oro Grande on a back road north of Victorville, about 185 highway miles west of Needles.

- The Toltec Lost Turquoise Mine: Said to be somewhere in the Crescent Mountains.

- The Treasure of San Timoteo Canyon: Hidden in San Timoteo Canyon near El Casco.

- The Treasure of Spanish Fort: Located near Fawnskin and Big Bear Lake in the San Bernardino National Forest.

- Wilson's Lost Mine: Said to be somewhere between the Providence and Old Woman Mountains near Danby.

San Diego County

- The Treasure of San Diego Mission: At or near Mission San Diego de Alcala, which is at 10818 San Diego Mission Road in the Mission Valley area of the city of San Diego.

- The Santa Ysabel Mission Treasure: In the foothills about 40 miles inland from the San Diego Mission.

- The Spanish Priest's Saddlebags Gold: Reported to have been hidden north of San Diego Mission on the "old road to Los Angeles," of which there are several, including what is now California Route 1, the Pacific (Pacific Coast) Highway.

- Antonio Joseph's Lost Placer: On a stream or stream bed in the Coyote Mountains of the Anza-Borrego State Park, near the Imperial County line.

- Bell's Lost Hideout Canyon Mine: In a canyon in the Vallecito Mountains.

- Bluebeard Watson's Buried Treasure: Somewhere in the Borrego Desert, probably in the Anza-Borrego Desert State Park, near the Imperial County line.

- Carmelita's Lost Gold: Said to be located on San Felipe Creek.

- Cortéz Treasure of San Luis Rey: Said to be in the vicinity of the San Luis Rey Mission in the town of the same name.

- Francisco de Ulloa's Treasure: Hidden somewhere in the San Luis Rey Valley near Oceanside.

- Joe Reedy's Lost Chicago Placer: On a stream or stream bed in the vicinity of what was once designated as Railroad Tunnel 21 in the Jacumba Mountains.

- The Lost Jacumba Treasure: Located near the Mexican border in the Jacumba Mountains, and possibly near the town of Jacumba.

- The Lost Portuguese Mine: Said to be somewhere in the Fish Mountains.

- Negro Jim's Phantom Mine: Thought to be in the vicinity of Yaqui Well.
- Old Hank's Lost Mine: Borrego Springs area, possibly in the Anza-Borrego Desert State Park.
- Sonora Joe's Lost Gold: Said to be in or near Fish Creek Canyon.
- The Lost Gold of Fish Canyon: Said to be in Fish Canyon, on the east fork of the San Gabriel River near Oceanside.
- The Black Crow Lost Mine: At the north end of Blair Valley.
- The Burnt Wagon Treasure of Borrego Springs: South of Borrego Springs in the Borrego Badlands area.
- The Fallbrook Stagecoach Treasure: Located between Temecula (Riverside County) and Pala on the old Butterfield Stage route, which is now County Route 16.
- The Gold Ledge of Vallecito Springs: Said to be located near Vallecito Springs.
- The Indian Gambler's Lost Placer: On a stream or stream bed in the vicinity of or in San Felipe Wash.
- The Indian Nuggets of the Vallecito Mountains: Said to be somewhere in the Vallecito Mountains.
- The Iron Door Cave Treasure of Ramona: In the Santa Maria Valley near Ramona.
- The Lost Gold in the Oriflamme Mountains: Hidden in the Oriflamme Mountains east of Julian.
- The Lost Indian Nuggets: In the Bucksnort Mountains near Collins Valley.
- The Lost Mine of Jesus Arroa: Somewhere in the Cocopah Mountains.
- The Lost Placer of Sentenac Canyon: On a stream or stream bed in Sentenac Canyon near Grapevine Mountain.
- The Lost Silver Mine of El Cajon Mountain (also known as the Lost Wadham and the Lost Barona): Located on or near El Cajon Mountain.
- The Lost Treasure at Russian Spring: Located near the old Russian Spring on outskirts of the city of San Diego.
- The San Felipe Stage Treasure: Said to be located in the

vicinity of the former site of the old San Felipe Stage Station in the San Felipe Valley.

- Vallecito Stage Station Treasure: In the canyon near the old Vallecito Stage Station (not to be confused with the town of Vallecito in Calaveras County).
- The Williams Lost Silver Mine: Located in the hills near Alpine.
- Joaquín Murietta's Treasures: During the 1850s, California's notorious bandit and murderer was active in and around nearly all the mining camps and towns in Calaveras, Fresno, Mariposa, San Diego, San Joaquin, Shasta, Tuolumne, and Ventura counties, and caches of his plunder are rumored to be hidden in each.

San Joaquin County

- The Treasure of Andrus Island: Located on Andrus Island in the San Joaquin River Delta near Isleton.
- Joaquín Murietta's Treasures: During the 1850s, California's notorious bandit and murderer was active in and around nearly all the mining camps and towns in Calaveras, Fresno, Mariposa, San Diego, San Joaquin, Shasta, Tuolumne, and Ventura counties, and caches of his plunder are rumored to be hidden in each.

San Luis Obispo County

- The Cave Landings Treasure: In a cave on or near the coastline in the vicinity of Avila Beach.
- The Chinaman Lost Mine: Somewhere in the La Panza Mountains.
- The Lost Mines and Treasure of the San Luis Obispo Mission: In the mountains near the Mission San Luis Obispo de Tolosa in San Luis Obispo.
- The Robber's Cave Treasure: In the cave known as the Robber's Cave, somewhere in San Luis Obispo County.

Santa Barbara County

- Mantz's Lost Coin Treasure: Said to be hidden near Ventucopa, probably in the Los Padres National Forest.

- The Lost Mine and Treasure of Santa Ynez Mission: In the mountains near the old Santa Ynez Mission.
- The Lost Mines and Treasure of Santa Barbara Mission: In the Santa Ynez Mountains near the city of Santa Barbara.
- The Pirates' Treasure of Arroyo Burro: Located in or near Arroyo Burro near Veronica Springs, west of Santa Barbara.
- The Treasures of San Marcos Pass: Said to be hidden on or near San Marcos Pass, which is located on Route 154 in the Santa Ynez Mountains.

Santa Clara County

- The French Saddlemaker's Treasure: Located in the vicinity of Santa Teresa Spring, south of San Jose.
- The Treasure of Rancho Teresa: On the site of Rancho Teresa in the city of San Jose.

Shasta County

- Alvey Boles's Lost Mine: Said to be located north of the historic mining town of Shasta (not to be confused with the Siskiyou County town of *Mount* Shasta), and possibly in the Whiskeyton-Shasta-Trinity National Recreation Area.
- Dutch Engle's Lost Mine: Said to be somewhere in the Arbuckle Mountains.
- Rattlesnake Dick's Lost Treasure: Hidden near Clear Creek in either Shasta or Trinity County.
- Old Man Waite's Pocket of Gold: Located on the Buckeye Ranch near Redding.
- The Bear Creek Cave Treasure: Located on Bear Creek north of the Shasta-Tehama County line.
- The Lost Cabin Mine of Soda Creek: Probably in the vicinity of an abandoned cabin or the ruins of a cabin near Soda Creek.
- The Lost Mine of Cow Creek: Located on Cow Creek near Bells Vista.
- The Lost Mormon Treasure of Clear Creek: Said to be in Clear Creek near Redding.

- The Rifle Barrel Payroll Treasure: Located in or near French Gulch.
- The Ruggles Brothers' Treasure: Located on Middle Creek, about 6 miles from Redding.
- Joaquín Murietta's Treasures: During the 1850s, California's notorious bandit and murderer was active in and around nearly all the mining camps and towns in Calaveras, Fresno, Mariposa, San Diego, San Joaquin, Shasta, Tuolumne, and Ventura counties, and caches of his plunder are rumored to be hidden in each.

Sierra County

- Lingard's Lost Lake of Gold: Said to be located in the Sierra Nevada near Downieville.
- The Lost Mine of Castle Ravine: In or near Castle Ravine near Downieville.
- The Lost Mine of Kanaka Creek Canyon: In or near Kanaka Creek Canyon.
- Thomas Stoddard's Lost Lake of Gold: A large quantity of gold, probably in small pieces, is reported to have been dumped in a lake, the name of which is not known, somewhere in the Sierra Nevada near Downieville.

Siskiyou County

- Anderson's Lost Waterfall Gold Pocket: Near or at a waterfall on the western slope of Mount Shasta.
- Hawkins's Lost Mine: Said to be located in the vicinity of McCloud on Route 89 near Snowman's Hill Summit.
- The Lassen Stage Treasure: Hidden somewhere on the western slope of Mount Shasta.
- The Lemurians's Treasure Vault: Said to be somewhere in the Cascade Mountains, northeast of the Sacramento River.

Sonora County

- The Spanish Mine of Sonora Pass: Located near the summit of Sonora Pass on Route 108 near the Mono County line.

Stanislaus County

- The Lost Gold Cache of Hawkins Bar: Said to be located in or near the old mining camp of Hawkins Bar.
- The Stanislaus River Flood Treasure: On the Stanislaus River flood plain, somewhere between Oakdale and Knights Ferry, probably buried beneath water level.
- The Teamster's Cache of Somes Bar: Said to have been hidden northeast of Somes Bar.
- The Treasure of Castle Crags: In the Castle Crags area near Dunsmuir.
- The Treasure of Trinity Mountain: On Trinity Mountain.

Tehama County

- The Lost Placer at Gold Bluff: Said to be located on a stream or stream bed near a place called Gold Bluff, near the town of Red Bluff on Interstate 5.

Trinity County

- Double Cabin Lost Mine: Located on the north fork of the Trinity River in the Salmon Mountains of the Shasta-Trinity National Forest.
- Lieutenant Jonas Wilson's Poker Cache: Said to be cached in Hoaglin Valley at the foot of Haman Ridge.
- Rattlesnake Dick's Lost Treasure: Hidden near Clear Creek in either Shasta or Trinity County.
- The Gold Cache of Canyon Creek: Hidden along Canyon Creek.
- The Gold of Hall City Cave: Said to be located in a cave in the vicinity of Weaverville on Route 299.
- Toby Bierce's Treasure: Said to have been buried in or near Cleveland Meadows (formerly known as Bierce Meadows).

Tulare County

- The Lost Haunted Mine: Located near Deer Mountain in the Sierra Nevada.

- The Lost Pipe Clay Mine: Said to be in the vicinity of what is now Sequoia National Park, possibly in the park or the Sequoia National Forest.
- Daniel's Lost Mine: Said to be located in the vicinity of the old mining settlement of Dogtown.

Tuolumne County

- Joe Thompson's Lost Gold: Located in the Big Creek Basin, south of Yosemite National Park in the Sierra National Forest.
- The Buried Treasure of Tom Davis: In or near the restored Gold Rush–era boomtown of Columbia, now designated as Columbia State Historic Park.
- The Dutchman's Lost Bonanza: Located near the above-described town of Columbia, it is probably identified for a German, or "Deutsche," man, many of whom were active in the area during the Gold Rush.
- The Lost Iron Kettle Treasure: According to legend, the "kettle" was buried about a half mile east of Columbia.
- The Tub of Coins Treasure: According to legend, the "tub" is buried at or near the old mining town of Yankee Hill near Sonora.
- The Lost Gold Ledge of Big Oak Flat: Said to be located in the vicinity of Big Oak Flat on Route 120.
- Joaquín Murietta's Treasures: During the 1850s, California's notorious bandit and murderer was active in and around nearly all the mining camps and towns in Calaveras, Fresno, Mariposa, San Diego, San Joaquin, Shasta, Tuolumne, and Ventura counties, and caches of his plunder are rumored to be hidden in each.

Ventura County

- Juan Chavez's Lost Treasures: Two different treasures have been attributed to him, one near Sage in Riverside County, and one on Agua Blanca Creek, near its confluence with Piru Creek in Ventura County.

- Old Pete's Lost Gold Mine: Located on Pine Creek south of Lebec (Kern County).
- The Airplane Crash Treasure on White Mountain: On White Mountain south of Gorman (Los Angeles County).
- The Lost Mine of Rayes Peak: Located on Rayes Peak near Ojai.
- The Lost Mine of Frazier Mountain: Said to be on Frazier Mountain southwest of Gorman (Los Angeles County).
- The Lost Treasure of Rancho San Miguel: Located on the old Rancho San Miguel, southeast of the city of Ventura.
- The Padre's Lost Mines: Located on Almo Mountain, west of Gorman (Los Angeles County) and west of old Fort Tejon (Kern County).
- Three Finger Jack's Treasure: Said to be located in the vicinity of Wheeler Hot Strings, north of Ojai in the Los Padres National Forest.
- Joaquín Murietta's Treasures: During the 1850s, California's notorious bandit and murderer was active in and around nearly all the mining camps and towns in Calaveras, Fresno, Mariposa, San Diego, San Joaquin, Shasta, Tuolumne, and Ventura counties, and caches of his plunder are rumored to be hidden in each.

Yolo County

- William Miner's Treasures: Said to have been hidden near Angels Camp in Calaveras County and near Woodland in Yolo County.

Yuba County

- Billy Snyder's Treasure: Believed to be located on Organ Creek near Comptonville.
- The Lost "River of Gold" Mine: With its exuberantly exaggerated appellation, this site is said to be located somewhere in the Marysville vicinity.

15

THE TREASURES OF THE NORTHWEST

Oregon

The Spanish visited the Oregon coast as early as 1543, but did not stay. The French, the Russians, British (including Captain James Cook), and Yankee sea captains from Boston all visited the Oregon coast in the eighteenth century, but they didn't stay. Lewis and Clark also visited, spent the winter, and then they went home. Soon after, however, John Jacob Astor set up his fur-trading post at Astoria.

The first people of European origin to put down roots in the Beaver State actually came for beaver rather than gold. But the fur trade *was* golden. This made the British jealous enough to seize Oregon, and the Americans became more jealous and seized it back.

The first mass migration into Oregon came with the people who struggled over the Oregon Trail to settle in the rich farmland of the Willamette River valley which snuggles between the coastal mountain ranges and the Cascades. The pot of gold that lay at the end of their Oregon rainbow was rich black dirt.

Nevertheless, when the California Gold Rush brought its cavalcade of prospectors into the West, some of them reasoned that the long veins of gold that seem to run north to south through the Sierra Nevada, also had sinews running north through the Cascades into Oregon. They were right. No one ever found gold deposits in Oregon to rival those of California, but there were sizable discoveries in the Rogue River country around Jackson and Josephine Counties, where place names like Gold Hill and Gold Beach remain as testaments to an exciting past.

Of the fifty states that we contacted for clarification of laws governing treasure hunting, Oregon was one of eleven that supplied useful information. The office of their attorney general replied that, under Oregon Revised Statutes 273.718-273.742, the state requires a permit for exploration for and removal of goods. The treasure trove program is administered by

Oregon Division of State Lands
775 Summer Street NE
Salem, Oregon 97310

Baker County

- The Lost Chinese Gold at Sparta: Near the old mining town of Sparta in the Wallowa Mountains, which is about 40 miles east of Baker City, most of the way on a paved road.
- The Snake River Treasures: Sites so designated are said to be located along the Snake River near Homestead in Baker County, and near Buckhorn Spring in Wallowa County.

Clackamas County

- The Blazed Cedar Treasure: The blazed cedar itself, if it could be found, would be the ideal clue to the cache, said to be located on Laurel Hill southeast of Portland.

Clatsop County

- The Coxcomb Hill Cache: Located on Coxcomb (or Cockscomb) Hill near Astoria.
- The Flavel House Cache: Said to be hidden at the house of the same name in Astoria.

- The Lost Mine of the Clatsop Plains: Said to be on or near the Clatsop Plains northeast of Seaside.
- The Lost Mine of Fishhawk Falls: Said to be at Fishhawk Falls, near Jewell on the Nehalem River.

Coos County

- The Randolph Trail Treasure: Hidden somewhere along the old Randolph Trail.

Crook County

- The Four Dutchmen's Lost Mine: Ochoco Mountains, Crook County.
- The Treasure at Skeleton Rock: The rock is said to be located in the vicinity of Prineville.

Curry County

- The Lost Soldier Mine: Said to be located in the Coquille Mountains, inland from Port Orford and possibly in the Grassy Knob Wilderness Area.
- The Treasures at Gold Beach: Located near the town of Gold Beach at the mouth of the Rogue River.

Douglas County

- The Two Frenchmen's Lost Cabin Mine: Said to be located in the vicinity of Illahee Peak.
- Edward Schieffelin's Lost Gold: One of his several rumored caches is said to be located near Days Creek or Coffee Creek (*see also* Jackson County).

Harney County

- The Virgin Gold of the Malheur River: Located near the Malheur River in the Malheur Mountains.

Hood River County

- Ben Smith's Mystery Mine: Said to be located in the vicinity of Mount Jefferson.
- The Lost Treasure of Horse Thief Meadows: Said to be located at or near the place known as Horse Thief Meadows.
- The Stagecoach Robber's Cabin Cache: Said to be hidden in the Cascade Mountains south of The Dalles.

Jackson County

- Ed Schieffelin's Lost Mine: One of his several caches is said to be in the vicinity of the Rogue River, possibly near the town of Rogue River on Interstate 5 (*see also* Douglas County).
- The Lost Indian Cinnabar Mine: Said to be located near the site of the old Jewett Ferry on the Rogue River.
- The Crazy Dutchman's Lost Mine: An eccentric German (a "Deutsche" man) is said to have had a mine located in the vicinity of Eagle Point.
- The Lost Cabin Mine of Steamboat Mountain: Probably located on Steamboat Mountain.
- The Lost Gold of Gold Hill: Hidden at or near Gold Hill, a town on the north side of the Rogue about 20 miles west of Medford.
- The Lost Saddlehorn Gold: Located on or near Foot Creek.
- The Wilson Brothers' Lost Cabin Mine: Probably near (but not too near) their cabin (if it could be found) in the Siskiyou Mountains south of Jacksonville, possibly near the Little Applegate River.

Josephine County

- The Lost Mine of Wolf Creek: Said to be located in the vicinity of town of Wolf Creek, probably on the creek of the same name, in northeastern Josephine County.
- Henry Smith's Treasure: Said to be somewhere on the old Six Bit Ranch, north of Wolf Creek and the town of the same name.

- Louis Belfil's Lost Ledge: Said to be located in the vicinity of Kirby.
- The Indian Council Treasure: Located on Buckhorn Mountain near Merlin west of Grants Pass.
- The Lost Badger Mine: Located near Murphy on the Applegate River south of Grants Pass.
- The Pine Puzzle Treasure: Located on Grave Creek near Sexton Mountain.

Klamath County

- The Lost Treasure of Spencer Creek: Located on Spencer Creek near Keno, about 10 miles southwest of Klamath Falls.
- Mr. Barham's Spiritualists Treasure: Said to have been hidden somewhere on the outskirts of Keno.
- Set-'Em-Op's Lost Cabin Mine: Located near Crater Lake, possibly in Crater Lake National Park or the Winema National Forest.
- The Potato Patch Cache: Said to have been hidden in what was once a potato patch, in the Swan Lake Valley.
- The Lost Pedro Mine: Said to be located in the vicinity of Diamond Peak.

Lake County

- The Lost Gold of Lost Forest: Said to be at the place known as the Lost Forest, northeast of town of Silver Lake near Table Rock or Hayes Butte.

Lane County

- Bohemia's Mystery Mine: Located near Steamboat Creek and probably near Bohemia Mountain in southeastern Lane County, but in an area accessible only from the town of Steamboat in Douglas County.

Lincoln County

- The Pirate's Treasure at Cascade Head: Said to be located in the vicinity of Cascade Head on the Pacific Coast near US 101.

- The Treasure of Newport: Said to have been hidden in or near the town of Newport, which is located at the mouth of the Yaquina River.

- The Treasure of Seal Rock: At or near Seal Rock on the Pacific Coast south of Newport on US 101.

Malheur County

- Sheepherder Victor Casmyer's Lost Ledge: Said to be located in the vicinity of Freezeout Mountain.

- The Blue Bucket Lost Placer: On a stream or stream bed near, or on the North Fork of Malheur River, north of the town of Juntura.

- The Indian Cave Treasure: Located near Ontario and possibly near the Snake River.

- The Squaw Cave Treasure: In a cave, possibly very close to Lake Owyhee, most of whose shoreline is reachable only on foot or by way of unpaved roads.

- The Wagon Tire Lost Mine: Located on Crooked Creek, which enters the Owyhee River north of Rome on US 95.

Marion County

- Captain Smith's Lost Ledge: Located on Mount Horeb, south of Elkhorn.

Morrow County

- The Treasure of Hinton Creek: Said to be located on the North Fork of Hinton Creek, east of Heppner.

Tillamook County

- The Pirate Treasure of Neahkahnie Mountain: Located on or near Neahkahnie Mountain near Manzanita on Nehalem Bay.
- The Treasures of Neahkahnie Mountain: Located on or near Neahkahnie Mountain, and possibly a variation on the previous entry.
- The Treasure of Cape Falcon: Said to be hidden at Cape Falcon, now located in Oswald West State Park.

Umatilla County

- The Baker Ranch Buried Treasure: Located on the site of the old Baker Ranch, near Birch Creek, south of Pilot Rock.
- The Lost Gold of Willow Springs: Located on the north slope of Battle Mountain, possibly near the state park of the same name near US 395.
- The Treasure of Stage Gulch: In Stage Gulch near Stanfield.

Wallowa County

- John Cash's Lost Mine: Said to be located in the vicinity of Lostine, probably in the Wallowa Mountains (the person referred to is not the popular country and western singer).
- The Lost Mine on Sacajawea Peak: Located on Sacajawea Peak in the Wallowa Mountains.
- The Lost Bear Creek Mine: Located near Bear Creek in the northeastern Wallowa Mountains.
- The Snake River Treasures: Sites so designated are said to be located along the Snake River near Homestead in Baker County, and near Buckhorn Spring in Wallowa County.

Washington County

- The Lost Gold of Gales Creek: Said to be located somewhere on Gales Creek.
- The Tillamook Burn Lost Mine: Said to be located within

or near a forest fire burn area on the Tillamook County line known locally as the "Tillamook Burn."

WASHINGTON

Until the middle of the twentieth century, when wartime urgency and cheap electricity from Columbia River dams turned Seattle into a world center of aerospace technology, the Evergreen State's gold was in its trees. Timber and timber products are still important, although they now share the balance sheet with the products of the companies started by Bill Boeing and Bill Gates.

Though some item of trivia may be found to bring the assertion into question, it can probably be said with reasonable certainty that Washington is the only mainland state west of the Rockies that never had a gold rush. There have been gold mines in the state, but no major deposits have been turned up to rival those modest discoveries in neighboring Idaho's Clearwater country and Oregon's Rogue River Valley.

There are stories of "lost Indian gold," and other lost gold in both the Cascades and the Olympics, but most of the treasure in these caches, if it is still cached, originated out of state. As with the states of the plains, Washington's treasure lore includes a lot of plunder. There are some especially important treasures rumored to be hidden around Port Townsend, and on the cool and quiet inlets of Willapa Bay.

Washington is also home to one of the twentieth century's most engaging "pirate" treasure stories, and one of the Northwest's most enduring folk legends. On November 24, 1971, a man identifying himself as "Dan Cooper" boarded Northwest Airlines flight 305 from Portland to Seattle. Shortly after takeoff, Cooper opened a briefcase filled with dynamite attached to a timer and requested $200,000 in twenty-dollar bills and four parachutes. In Seattle, his demands were met, the passengers were released, and the Boeing 727 took off again with only its flight crew aboard.

The jetliner was flying at 200 mph and was 10,000 feet above the Lewis River area of southwest Washington when the flight crew felt a jerk as one of the rear doors came open.

When they looked, the door was in fact open, and Cooper, the parachutes, and the money were gone.

Identified as "D. B. Cooper," although he didn't actually call himself that, the skyjacker was never found, nor was most of the money. Over the ensuing months, rumors of sightings rivaled those of Elvis Presley two decades later, but no one ever saw the parachutes come down and there were no confirmed sightings of a man with suitcases full of money. Folk songs were written and tales were spun, but no solid leads were ever found until 1980, when $5,800 was discovered on the banks of the Columbia River near Vancouver, Washington.

The woods remain silent. Maybe Cooper travelled to some tropical island to live out his years in luxury, or maybe he died in his plunge from that Northwest 727. Maybe the $194,200 was spent years ago, or maybe it deteriorated in a pulpy mass under the silt on the bottom of the Lewis or Columbia River, or maybe it's buried under tons of debris from the Mount St. Helens eruption of 1980. Or maybe it lies in a relatively dry place, protected from the elements and awaiting discovery. Many have bet on the latter, and many have been disappointed.

Asotin County

- The Lost Shovel Creek Mine: Said to be located on or near Shovel Creek near where the Grande Ronde River flows into the Snake Rivers.
- The Prospector's Lost Treasure: Said to be near Rogersburg, a town on the Snake River accessible only on an unpaved road.
- The Trio Lost Mine: As with the previous entry, this one is located near Rogersburg.
- The Treasure of the Grande Ronde River: Located on the Grande Ronde River, about 35 miles south of Clarkston.
- The Water Pail Lost Mine: Said to be located in the vicinity of Anatone.

Clallam County

- Victor Smith's Lost Treasure: Located near Port Angeles on the north side of the Olympic Peninsula, possibly in the Olympic National Park.

- The Chevy Chase Treasure: Said to be hidden on or near Port Discovery Bay.

- Chief Sam Elwtha's Lost Gold Cave: The cave is somewhere in the Olympic Mountains (possibly on Hurricane Ridge) near the Elwtha River, which originates near Mount Seattle and flows into the Straight of Juan de Fuca about 5 miles west of Port Angeles.

Clark County

- D. B. Cooper's Lost Loot: $194,200 in $20 bills, possibly lying somewhere in the woods of the Lewis River area.

Cowlitz County

- The Lost Mine of Ostrander Creek: Said to be on or near Ostrander Creek.

Douglas County

- The Karr Treasure: Said to be located in the vicinity of Bridgeport.

Grant County

- John Welch's Lost Gold Placer: On a stream or stream bed somewhere in the vicinity of Trinidad near the Columbia River.

Grays Harbor County

- C. R. Terrell's Lost Placer: On a stream or stream bed in the Olympic Mountains north of Montesano.

Jefferson County

- The Mountain Peak of Gold: A fanciful story of an impossible natural feature, this story may have some basis in a factual place, somewhere west of Brinnon in the Olympic Mountains of Olympic National Park.
- The Indian Treasure: Said to be located in the vicinity of Point Ludlow, possibly on Quinault land.
- The Buckley Treasure: Said to be in or near the city of Port Townsend.
- The Treasure of Discovery Bay: Located in the vicinity of the town of Discovery Bay on US 101 about 25 miles southwest of Port Townsend, Jefferson County.
- Harry Sutton's Lost Treasure: Said to be located in the vicinity of Port Townsend.

King County

- Lars Hanson's Lost Treasure: Said to be hidden somewhere on Vashon Island in Puget Sound.

Kittitas County

- The Lost Nuggets of Swauk Creek: Located somewhere along Swauk Creek near Cle Elum.
- Chief Kitsap's Lost Gold Mine: Said to be located east of Ellensburg in the vicinity of the Greenwater River.

Klickitat County

- The Lost Mine of the Klickitat: Said to be located in the vicinity of Glendale.

Okanagan County

- Chief Smitkin's Lost Treasure: Located near St. Mary's Indian Mission, near Omak, possibly on the Colville Indian Reservation.

Pacific County

- The Treasure of Long Island: On Long Island, now designated as Willapa National Wildlife Refuge, in Willapa Bay near the mouth of the Naselle River.
- Captain James Scarborough's Cache: Said to have been stashed in or near Chinook or the Lewis and Clark Campsite State Park, at the mouth of the Columbia River.
- Captain James Johnson's Treasure: Hidden somewhere on the old Johnson homesite in Ilwaco near Cape Disappointment.

Skamania County

- The Lost Spanish Mine: Said to be in Gifford Pinchot National Forest, somewhere in the Cascade Mountains between Mount Adams and Mount St. Helens.

On or near the border of Skamania and Yakima Counties

- Pierre Rabado's Lost Gold Mine: Located near Mount Adams (Yakima County) in the Cascade Range, possibly in the Mount Adams Wilderness Area of the Gifford Pinchot National Forest, or on the Yakima Indian Reservation.

Stevens County

- The Highgrader's Poor Farm Treasure: The term "highgrader" implies fraud, and "poor farm" implies little value, but for what it's worth, the site is said to be located in the vicinity of Colville, possibly in the Colville National Forest and possibly on state-owned or formerly state-owned land.

• The Robber's Roost Treasure: Located at the place known as "Robber's Roost," in the vicinity of Fruitland near the Columbia River.

Walla Walla County

• The Lost Mine of Jarbow Meadows: Said to be located in the vicinity of Jarbow Meadows in the McIntyre Range.
• The Walla Walla Cache: Located near Wallula on US 730 near the Columbia River, about 35 miles west of the city of Walla Walla.

Whitman County

• The Blackberry Treasure: Said to be located in the vicinity of Oakesdale.
• The Blue Lady Treasure: Located in the vicinity of Steptoe on US 195, possibly near Steptoe Butte State Park.

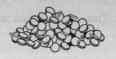

16

THE TREASURES OF ALASKA AND HAWAII

ALASKA

Even if the details have faded from our minds, we all remember from our elementary school American History that when Secretary of State William Henry Seward bought Alaska from the Russians in 1867, folks called the place "Seward's Folly" because they thought that he'd been swindled by the wily Tsar Alexander II. We also remember the punch line, which was that the $7.2 million that Seward paid (that's two cents an acre) was a fabulous bargain when measured against the riches that Alaska has poured into the United States economy.

Today we think of Alaska's economy as a tenuous conflict between North Slope oil and ecotourism, but during its first century, Alaska enjoyed a series of booms based on furs, whaling, and gold. Indeed, gold alone earned back the purchase price of Alaska fifty times over before 1910. Gold discoveries were made throughout the 1890s, especially around Fairbanks and Nome, as well as in the southeast, adjoining the Klondike region of Canada's Yukon Territory. The 1896–1898 Klondike Gold Rush was the last of the great nineteenth century gold rushes, and at its peak in the last year of the century, Klondike claims grossed six *billion* dollars (in 1990s dollars).

In the early days, Alaska gold was placer gold, close to the surface and relatively easy to get. Naturally, placer mining reached the point of exhaustion within time, and the miners eventually abandoned their claims. As the gold became harder to reach, Alaska's harsh climate and difficult terrain made looking for it less economically viable, and people gave up. As with other states—from Georgia to California—where there have been gold rushes, stories still persist of lost mines and old cabins with incredible caches.

Despite advances in the technology in everything from thermal insulation for gloves to air travel, Alaska's environment is essentially no kinder to the interloper today than it was in 1898. Outside the three or four major cities, and a handful of corridors with paved roads, Alaska is a true wilderness, where moving about and existing in the vast and literally trackless terrain is complicated by darkness and sub-zero cold in winter and mosquitos "the size of horses" in summer. With this in mind, we have chosen to recommend against treasure hunting in Alaska. We list two sites as points of interest, of which the Anvil Creek area is easily accessible from Nome and potentially rewarding. For those people wishing to visit Alaska to get the feel of the Gold Rush era, the Klondike Gold Rush National Historical Park in Skagway is a good place to start, and an excursion into the Yukon via the White Pass and Yukon Railroad is an excellent way to continue.

Nome County

- The Beaches of Nome: The site of a huge gold rush in September 1898, the beaches are still a paradise for not only amateur gold hunters, but other treasure and bottle hunters as well, especially around Anvil Creek, about 4 miles north of Nome.

Valdez-Cordova County

- The Lake of the Golden Bar: About 500 pounds of nuggets were hidden in a lake near a mining site somewhere in the St. Elias Mountains near the Yellow River and the Canadian border.

HAWAII

While most of the United States was once claimed by monarchies, and several states were once well-developed colonies of monarchies, Hawaii is the only state that once *was* an independent monarchy. This came to an end in 1893, when Queen Liliuokalani tried to abolish the constitution. The people, which by this time included a large number of American ex-patriots, revolted and established a republic that survived until 1898 when Hawaii was annexed as a territory of the United States.

With a royal history comes the intrigue that often swirls around monarchies, and of course, the material wealth, the literal and figurative "crown jewels" that always accompany monarchies. The seat of Hawaii's royals was on the island of Oahu, which is now the most densely populated of the five "main" islands.

Despite the incredible density of population on Oahu's southern shore, which includes the Honolulu metropolitan area, rumors of hidden treasure yet persist. There are beaches on Oahu that are still remote, and the steep slopes of the rugged Koolau Mountains that run the length of the eastern side of the island offer many hiding places. Ironically, one of Hawaii's most amazing legendary caches is within a few minutes walk of crowded Waikiki Beach. As the story goes, $100,000 to $150,000 worth of five-sided gold coins minted by the Hawaiian monarchy are said to lie buried in a cave in the Aina Moana State Recreation Area, immediately west of Waikiki and not far from Ala Moana Boulevard.

Other treasure sites on Hawaii's islands are more remote and center on stories of pirate loot. While pirates and plunderers did visit Hawaii's secluded coves and black sand beaches, there are much fewer sites here than on the eastern seaboard of the continental United States because Hawaii was much farther from major trade routes. With the possible exception of the California Gold Rush, the huge volume of gold and other valuables that passed through the Caribbean and coasted up to New England in the seventeenth and eighteenth centuries was many time greater than the value of commerce anywhere else in the Western Hemisphere or most of the Pacific until the twentieth century. Hawaii was an important stopping place for trans-Pacific shipping during that time, but it was isolated from the major ports of Asia and North America by a radius of over 2,000 miles, a distance that was measured in weeks rather than hours, as with points in the Caribbean and the eastern seaboard.

On the other hand, the process of searching for buried treasure at a remote beach on Molokai is always a much more pleasant undertaking than doing the same in Death Valley or the South Dakota Badlands. It just may be a bit less rewarding financially.

City and County of Honolulu (Island of Oahu)

- Kaena Point Pirate's Treasure: In 1823, six chests of pirate's treasure were hidden near some fitted stone walls at the top of Kaena Point.
- The Cave of Kings Treasure: Hidden somewhere on Ford Island in Pearl Harbor, now part of the US Navy's Pearl Harbor Naval Reservation.
- The Keakauailau Treasure: A large hoard of gold is said to have been buried near the head of Moana Stream in the Koolau Range.
- The Hawaiian Monarchy Gold: $100,000 to $150,000 in coins are said to be buried in the Aina Moana State Recreation Area.

Hawaii County (Island of Hawaii)

- Captain Cavendish's Treasure: The pirate's cache of gold and silver is rumored to be buried at Palemano Point, on the west coast ("Kona Side") of Hawaii.
- Captain Turner's Treasure: The famous pirate is said to have buried a number of chests filled with gold and silver coins as well as church treasures in caves on the north side of Kealakekua Bay, possibly in or near Kealakekua State Park or Hikiau Heiau State Monument.

THE PHENOMENAL
NATIONAL BESTSELLERS
FROM TRACY KIDDER

A·M·O·N·G
SCHOOLCHILDREN

71089-7/$12.00 US/$16.00 Can

For an entire year Tracy Kidder lived among twenty schoolchildren and their indomitable, compassionate teacher—sharing their joys, their catastrophes, and their small but essential triumphs.

The SOUL OF A NEW MACHINE

71115-X/$12.50 US/$16.50 Can

Tracy Kidder's "true life-adventure is the story of Data General Corporation's race to design and build the Eagle, a brand new 32-bit supermini computer, in the course of just a year and a half…compelling entertainment."

Washington Post Book World

HOUSE

71114-1/$12.50 US/$16.50 Can

With all the excitement and drama of a great novel, Kidder now takes us to the heart of the American dream—into the intimate lives of a family building their first house.

Astonishing UFO Reports
from Avon Books

COMMUNION: A TRUE STORY
by Whitley Strieber   70388-2/$6.99 US/$8.99 Can

TRANSFORMATION: THE BREAKTHROUGH
by Whitley Strieber   70535-4/$4.95 US/$5.95 Can

THE GULF BREEZE SIGHTINGS
by Ed Walters and Frances Walters
70870-1/$6.99 US/$8.99 Can

UFO ABDUCTIONS IN GULF BREEZE
by Ed Walters and Frances Walters
77333-3/$4.99 US/$5.99 Can

THE UFO CRASH AT ROSWELL
by Kevin D. Randle and Donald R. Schmitt
76196-3/$5.99 US/$7.99 Can

THE TRUTH ABOUT THE UFO CRASH AT ROSWELL
by Kevin D. Randle and Donald R. Schmitt
77803-3/$6.99 US/$8.99 Can

A HISTORY OF UFO CRASHES
by Kevin D. Randle   77666-9/$5.99 US/$7.99 Can